TELEPHONY, THE INTERNET, AND THE MEDIA

*Selected Papers from the
1997 Telecommunications
Policy Research Conference*

TELECOMMUNICATIONS

A Series of Volumes Edited
by Christopher H. Sterling

TELEPHONY, THE INTERNET, AND THE MEDIA

*Selected Papers from the
1997 Telecommunications
Policy Research Conference*

Edited by

Jeffrey K. MacKie-Mason
University of Michigan

David Waterman
Indiana University

LEA   LAWRENCE ERLBAUM ASSOCIATES, PUBLISHERS
1998   Mahwah, New Jersey   London

Lawrence Erlbaum Associates, Inc., Publishers
10 Industrial Avenue
Mahwah, New Jersey 07430

Cover design by Jennifer Sterling

Library of Congress Cataloging-in-Publication Data

Telecommunications Policy Research Conference (25th: 1997)
    Telephony, The Internet, and the Media: Selected Papers from
    the 1997 Telecommunications Policy Research Conference / edited by
    Jeffrey K. MacKie-Mason, David Waterman.
              p.   cm.
    Includes bibliographical references and index.
    ISBN 0-8058-3151-7 (hardcover: alk. paper)—ISBN 0-8058-3152-5 (pbk: alk. paper)
    1. Telecommunication policy—Congresses. 2. Telephone—
Congresses. 3. Internet (Computer network)—Congresses.
    4. Mass media—Congresses.
    I. MacKie-Mason, Jeffrey K. II. Waterman, David. III. Title.
HE7645.T45   1997
384—dc21   98-25263
                                              CIP

Books published by Lawrence Erlbaum Associates are printed on acid-free paper, and their bindings are chosen for strength and durability.

Printed in the United States of America
10 9 8 7 6 5 4 3 2 1

Contents

Authors

Dan L. Burk
Seton Hall University

Angela J. Campbell
Georgetown University Law Center

David D. Clark
Massachusetts Institute of Technology

Lorrie Faith Cranor
AT&T Labs-Research

Douglas Galbi
Federal Communications Commission

Farid Gasmi
*IDEI, Université de Toulouse I
    and Université de Bretagne Occidentale,
    Brest, France*

Willie Grieve
Saskatoon, Saskatchewan

Robert B. Horwitz
University of California, San Diego

Heather E. Hudson
University of San Francisco

Jean-Jacques Laffont
IDEI, Université de Toulouse I, France

Brett A. Leida
Massachusetts Institute of Technology

Stanford L. Levin
*Southern Illinois University
    at Edwardsville*

Jeffrey K. MacKie-Mason
University of Michigan

Lee W. McKnight
Massachusetts Institute of Technology

Judith A. Molka-Danielsen
Molde College

Bruce M. Owen
Economists Incorporated

Joseph Reagle, Jr.
*Massachusetts Institute of Technology,
    World Wide Web Consortium*

Mark Scanlan
*Consultant Economist,
    European Commission*

William W. Sharkey
Federal Communications Commission

Howard A. Shelanski
*University of California at Berkeley,
    School of Law*

David W. Sosa
University of California at Davis

David Waterman
Indiana University

Martin B. H. Weiss
University of Pittsburgh

Acknowledgments

The Telecommunications Policy Research Conference (TPRC) has now run for 25 years. Throughout this history, many of the papers presented have resulted in journal articles and other publications. For most of the years, there was not a regular proceedings publication.

In 1994, the combined efforts of Christopher H. Sterling, the series editor of the Lawrence Erlbaum Associates Telecommunications Series, Hollis Heimbouch of LEA, and TPRC Board members John Haring, Bridger Mitchell, and Jerry Brock, resulted in a new arrangement for LEA to publish a series of selected papers from the annual TPRC. This is the fourth volume in the LEA series, and it contains papers selected from the 25th annual conference, held in September 1997.

We greatly appreciate the assistance we received in dealing with the difficult task of selecting papers for this volume. We consulted with the TPRC organizing committee, session chairs and discussants, and various others who volunteered their time to referee. There were over 70 papers presented at the conference, and there were many outstanding contributions that we regret we could not include due to space limitations.

We have worked to fulfill our promise to deliver a published text in time for the 26th annual conference. We could not have done this without superb production work. We thank Cyndi Connelley, Roxie Glaze, and Jennifer Sterling for their expert copyediting entry and typesetting work. We are especially grateful to Linda Bathgate, Barbara Wieghaus, and their colleagues at LEA for guiding the production of this book to its conclusion.

We owe special thanks to the John R. and Mary Markle Foundation. Without its long running grant support, the 25th annual TPRC would not have been possible. In addition, the 25th TPRC had the benefit of a generous matching grant from the W. K. Kellogg Foundation. AT&T, Bell Atlantic, the Benton Foundation, and Cablevision Systems Development Corporation also made generous donations to the 1997 TPRC.

Jeffrey K. MacKie-Mason
David Waterman

Preface

My, what a long strange trip it's been! The first Telecommunications Policy Research Conference was held in 1972. In 1997 we celebrated the 25th annual conference. The attendees were an impressive group, but I hazard a guess that few of them would have forecast that in 25 years we would be watching AT&T struggle to get *back into* local telephony! Of course, *plus ça change, plus la même chose*, as Bruce Owen reminds us in his retrospective essay. Still, it is hard to imagine anyone involved in telecom policy over the last two and a half decades who does not find the experience just a bit astonishing.

It is conventional for a conference organizer to express gratitude for the opportunity. In this instance, my gratitude for the honor of being selected as the Organizing Chair of the 25th TPRC is genuine and deep. TPRC is a venerable institution (especially counted in Internet years). Along with telecommunications policy, it has grown and changed in unexpected ways. I'm sure the parents shook their heads with some puzzlement as the teenager grew into an adult. But one of the especially gratifying events this year was to gather some of those parents for the opening plenary, and to see the persistence and success of their original vision. We had a gathering of three or four generations of telecom policy family this year. As with any extended family, there were some disagreements. Nonetheless, I think we almost universally share an appreciation for this community of scholars, policymakers, and practitioners who are joined by a shared interest in telecom policy, despite widely ranging academic fields and organizational interests. I thank past conference organizers for inviting me to participate, and I thank last year's Board of Directors for honoring me with the Chairship.

Being chair was also (usually!) a delight. For this I must thank the members of the Organizing Committee: Jorge Schement of the Pennsylvania State University (and the Program Chair for the 26th Annual Conference); Paul Resnick of the University of Michigan; Jessica Litman of the Wayne State School of Law; Ben Compaine of the Pennsylvania State University; Heather Hudson of the University of

California, San Francisco; Evan Kwerel of the Federal Communications Commission; Nicholas Economides of New York University; Jean-Paul Simon of France Telecom. This team managed the challenging issues of forming a program for a conference that becomes more diverse each year. They bore with grace and generosity the burdens imposed by a significant increase in the size of the program to celebrate the 25th annual conference. The Committee also debated and implemented some significant policy changes, notably the requirement that completed papers be received before the conference begins. On the awkward task of enforcing this policy and on other difficult decisions, the Committee always reached consensus and acted with respect for authors combined with commitment to the success of the conference as a whole.

On behalf of the Committee, we would like to thank the nearly 200 scholars who submitted abstracts and papers in response to the conference call. Submissions increased by about 30% over the previous year. The most important determinant of conference quality is, of course, the quality of the papers presented. We were sad that we could not include every good paper, but we were pleased to encounter this problem.

On behalf of everyone participating in the conference, I want to offer heartfelt appreciation for the tremendous effort by Dawn Higgins and Lori Rodriguez, the conference administrators. It is clear to me each year that participants value Dawn and Lori; it is even more clear to me now that participants don't know the half of it. In addition, we are all grateful for the largely unheralded work by the Board of Directors, and especially Pam Samuelson and Eli Noam, who co-chaired last year's Board. The Board takes responsibility for financing the follies of the Organizing Committee; it also handles all of the difficult policy questions. Indeed, I made sure that the name of the Organizing Committee was changed to the Program Committee for the future because most of the important organizing work is done by Dawn, Lori, and the Board; the (now) Program Committee is largely left to focus on the fun part: putting together a first-rate program. For this we are grateful.

Let me close by thanking the authors of the papers selected for this volume. The selection was difficult: We chose 15 papers from among nearly 70 presented at the conference. The authors responded quickly and gracefully to the comments that David and I provided. On a personal note, I wish to thank David for his leadership in the editing of this volume. David played parent to my irresponsible child; he guided our selection of papers and preparation of introductory materials with a wise hand, and he did all this with good cheer. Now, we hope you, the reader, enjoy this collection of current, high-quality telecom policy research. And we look forward to seeing you at future TPRCs.

—*Jeffrey MacKie-Mason*

Introduction

Jeffrey MacKie-Mason
University of Michigan

David Waterman
Indiana University

With our title, *Telephony, the Internet, and the Media*, we intend to reflect the diversity of the Telecommunications Policy Research Conference. We also have more substantive reasons to group papers involving these seemingly different industries under the same roof.

One reason is the increasing difficulty of considering the policy problems of one communications industry in isolation from the others. Thirty years ago, probably the most significant relationship between telephony and the media was the cable industry's need for access to telephone poles to hang coaxial cable. Today, telephone companies and cable companies increasingly are offering each other's services to consumers, creating a maze of policy dilemmas that in the United States culminated in the Telecommunications Act of 1996. With the exceedingly fast growth of the Internet, from essentially nothing only 5 years ago, a third "industry" has complicated communications policy still further. On the input side, the Internet uses both telephone and cable facilities. The Internet's outputs include new communication forms that sometimes substitute for, sometimes complement traditional telephone service and media programs. Radio, newspapers, voice telephony, and streaming video all are found in various stages of maturity on the Internet. It is not reaching to say that no longer can any telecommunications policy be limited strictly to a single communications medium.

A second reason to combine papers on various communications industries is the opportunity to learn common policy lessons. Most communications industries share fundamental economic characteristics, such as strong economies of scale with respect to the number of consumers, and common social concerns, such as freedom of speech and privacy issues. As the chapters in this volume indicate, these common concerns are especially important for Internet research. Internet content control, standard setting, and network interconnection pricing, for example, parallel the same issues in the media or telephony. If policy learning for the Internet is to progress in the same blindingly fast "Internet years" that the medium itself is growing, it will be important to learn from experience in the media and telephony.

The remarkable pace of change that is merging the technological, economic, and policy issues in communications has continued unabated in the year since the last edition of the TPRC Selected Papers volume. There is no shortage of subject matter for TPRC authors! In the United States, implementation of the Telecommunications Act of 1996 has continued to unfold. Some effects of the Communications Decency Act are emerging slowly, partly due to still-pending legal challenges to provisions for local telephone company entry into long distance service, and to the pending Supreme Court case on the FCC interconnection rules. The debate on Internet content control was highlighted by the Supreme Court nullification of the Act's content "decency" provisions. In major developments, the FCC implemented telephony access charge reform and issued rules to implement universal access reform. Notable in the latter rules was the convergence between telephony and Internet policy: A nonprofit corporation was formed to establish universal Internet access through schools and libraries. In addition, most states have been adjusting their policies to conform to the new federal law.

1997 was a big year for international policy. The World Trade Organization reached a major agreement on telecommunications competition, in which 69 countries, representing over 90% of world telecommunications activity, agreed on a framework for far-reaching liberalization. Some of the consequences are discussed in this book. There was also a major international debate about the governance of the Internet, with a focus on the ownership and allocation of property rights in Internet address names. An ad hoc international management authority was established, but it remains to be seen whether it will be considered authoritative.

Despite the heavy attention to telephony and the Internet over the past couple of years, major developments continued in media policy. Notable in the United States were the promulgation of new guidelines for children's broadcast television, and mandated standards for a V-chip (a hardware device that allows parents to block access to programs carrying certain labels). The FCC also finalized its rules for allocating digital TV bandwidth to existing broadcasters, with a provision for auctioning reclaimed analog bandwidth in 10 years.

The telecom industries were busy while domestic and international policies were being revised. MCI agreed to be purchased by Worldcom for $30 billion in the largest U.S. deal ever. This dealwould combine the second and fourth largest U.S. long distance providers, and two of the leading Internet providers. SBC and Pacific Telesis completed their $16.5 billion merger in April; Bell Atlantic and NYNEX completed their $25.6 billion merger in August. New alliances among various national monopoly carriers were announced, in part anticipating the opening of cross-border telecom competition in the European Union.

With the continuing convergence and interaction between traditionally separate telecom industries, it is not easy to segment our 15 selected papers from the 1997 TPRC into well-defined categories. To assist the reader, we have organized the book into 5 parts. Part I is labeled Historical, a category that is special to the 25th anniversary edition of the TPRC. Part II, Telephony, Part III, The Media, and Part IV, The Internet, follow, and we conclude with Part V, Comparative Studies in Telephony and Satellite Policy. Readers will notice repeated themes and cross-connections between the chapters in these sections.

I. HISTORICAL

Bruce Owen, in "A Novel Conference: The Origins of TPRC" (Chapter 1), sets the context by reviewing the TPRC's 25 year history and its contributions to telecommunications research and policy making. Owen describes the Washington in which the first conference was held in 1972, under auspices of the old Office of Telecommunications Policy, as a "lonely and inhospitable place" to the academically minded in telecommunications. In making its journey from that 15-presenter meeting held at the old Executive Building, to the far larger present-day conference, the TPRC has left a substantial wake.

In reviewing the TPRC's contributions, Owen cites both its "inputs," an extraordinary increase—from virtually zero—in the number of economists and other professionals employed by the FCC and other Washington agencies involved with communications policy, and its "outputs," the TPRC's role in the revolutionary reforms in telecommunications regulation. The FCC, he notes, now routinely considers economic welfare effects in its deliberations, and a whole new accompanying set of arguments frame the Washington debate on telecommunications policy. The unique contribution of the TPRC, Owen notes in conclusion, is the "invisible college" of communication researchers throughout the world that collaborate and deliver relevant policy analysis to government agencies.

II. TELEPHONY

Recent U.S. telecom policy discussion has been dominated by the Telecommunications Act of 1996. This sweeping legislation was both a reaction to the major transformations in telecom technology and markets, and a (partial) roadmap to the future competitive landscape. Within this context, we have four chapters devoted to current issues in telephony. The first two concern regulation and competition in local service, whereas the next pair examine regulatory arbitrage pressures on international phone traffic.

In "A Technico-Economic Methodology for the Analysis of Local Telephone Markets" (Chapter 2), Farid Gasmi, Jean-Jacques Laffont, and William Sharkey develop a framework for modeling regulation that combines forward-looking cost analysis with the modern theory of regulation under asymmetric information. The authors present both method and results in an approach that combines an engineering process model of telecommunications service costs with an economic model of regulation and competition.

Gasmi et al. use their model to examine several long-standing issues in telecom regulation. In one such analysis they find that despite the problem of private information not directly available to regulators, the deviation between first-best and optimal regulated prices need not be large. They also find that the optimal regulatory prices can be well approximated through reasonably simple linear pricing rules. The authors then apply the method to a comparison of alternative regulatory approaches. They present many interesting results, including support for the superior performance of price cap regulation.

Local service regulation in the United States and elsewhere must increasingly accommodate facilities-based competition. In the United Kingdom, for example, most customers already have a choice between at least two providers. In the United States most wireline competition thus far has been through reselling (although wireless competition has been strong for over a decade), but cable providers and other overlay builders are preparing to offer competitive service.

Judith Molka-Danielsen and Martin Weiss, authors of "Firm Interaction and the Expected Price for Access" (Chapter 3), use a model related to that of Gasmi et al. to assess the effects of local competition on access pricing and universal service. They model duopoly pricing between two firms with no subsidy for universal service. Molka-Danielsen and Weiss then calibrate their cost and demand functions to proprietary data from multiple service areas. Without the cross-subsidy, penetration rates fall. The authors characterize the sensitivity of access prices and penetration to cost and demand conditions, as well as to the nature of strategic interaction between the two firms. One general conclusion is that fixed costs are sufficiently high that Bertrand (marginal cost pricing) equilibria are unlikely with only two facilities-based competitors.

The next two chapters examine consequences of the ad hoc system of accounting-based international "settlement" rates in a world where domestic deregulation and technological advance foster competitive adaptation. When regulatory price structures do not reflect cost or efficient bargaining outcomes, innovative service providers will seek ways to capture some of the regulatory inefficiency rents left on the table. In both of these chapters, we see that the market pressures leading to domestic telecom regulation reform throughout the world over the past two decades are now squarely challenging the framework of transnational regulation.

Douglas Galbi studies the consequences of regulatory by-pass opportunities in "The Implications of By-Pass for Traditional International Interconnection" (Chapter 4). International voice transit is treated as a jointly provided service in most bilateral treaties, and the revenues are shared according to a fixed, arbitrary rule. Because the revenue share may be above or below a competitive return on service, countries with multiple international providers, such as the United States, also have rules specifying the sharing of traffic volume among competing providers. However, following increased liberalization of domestic competition, unregulated alternative transit will now be permitted by 52 countries through the WTO agreement. Other by-pass opportunities also exist. Galbi models the pricing and traffic volume strategies for competing carriers who can choose between settlements traffic and by-pass. Just as we have seen for domestic by-pass in the United States following AT&T's divestiture, regulated and by-pass traffic can co-exist in equilibrium. However, Galbi shows that by-pass opportunities impose dynamic and complex constraints on the policy effectiveness of internationally regulated rates. It appears unlikely that fixed accounting policies can implement desired policy outcomes in a world of increasingly dynamic and multilateral unregulated competition.

In the final chapter of Part II, Mark Scanlan offers some surprising insights on the consequences of international regulatory arbitrage in "Call-Back and the Proportionate Return Rule" (Chapter 5). Call-back is a scheme to arbitrage artificial differences in international call origination prices. This type of arbitrage does not by-pass international settlement rates; rather, it reroutes calls to take advantage of different termination rates. Suppose that a French–U.S. call is priced higher when it originates in France. A call-back service provides French callers with a special U.S. number to call. The system extracts the originating French number, and originates a circuit from the U.S. side from which the French caller can reach its U.S. party.

Scanlan reports that in 1996, 42 of 66 responding countries had declared call-back services to be illegal. However, he shows that in most countries the provision of a call-back service can increase the profits of the operator in the higher priced country. A key feature of the argument straightforwardly illustrates how ad hoc pricing rules can stand intuition on its head: Because international connection revenues are fixed and split according to formula, it doesn't matter to the operator who originates the call. But if lower end-user prices stimulate demand, the high-cost operator earns a windfall. Together with Galbi's chapter, Scanlan's analysis suggests that the pressures for international rate reform in the wake of domestic reform will be great.

III. THE MEDIA

The first two chapters in this group address content regulation in television. Content regulation will probably always be with us; it certainly continues as a major focus for telecom policy. Although one of the major new stories of the year was the Supreme Court's decision striking down the Communications Decency Act as unconstitutional, content regulation in television is on the upswing after its near total eclipse during the broadcast deregulation era of the 1980s. Pursuant to the 1990 Children's Television Act, FCC license renewal guidelines that require stations to air 3 hours of children's educational television per week took effect in September 1997. Following the 1996 Telecommunications Act, the FCC has now approved the television industry's new program rating system and issued technical rules for installation of V-chips in TV sets. The television content debates have important implications for attempts to regulate Internet competition as well.

Howard Shelanski, in "Video Competition and the Public Interest Debate" (Chapter 6), takes a broad legal and economic perspective on content regulation. His main idea is that the traditional economic "market failure" arguments—spectrum scarcity, a lack of sufficient competition, limited channel capacity, and a lack of direct payment mechanisms—are outmoded and no longer justify government content intervention. After reviewing the history of FCC content regulation since the 1930s, Shelanski argues that explosions in the capacity and competition of video media, including pay television and videocassettes, render these arguments irrelevant. However, the participants in current content regulation debates, such as those involving children's television, continue to rely on the traditional arguments.

Shelanski then points out that broadcast content regulations, although they cannot be justified in economic terms, might be justified in terms such as whether parental preferences can and should replace those of children because the children are not equipped to be gatekeepers. He makes the case that this and other noneconomic arguments should be "unbundled" from the no longer appropriate arguments that regulation is needed to enhance program diversity or remedy gaps due to economic inefficiency in privately supplied programming.

Angela Campbell, in "Lessons from Oz: Quantitative Guidelines for Children's Educational Television" (Chapter 7), turns to the specific issue of whether the FCC's children's television guidelines are likely to work, and how they might be improved. She does so by examining Australia's long experience with a children's television quota. That experience, she argues, suggests that quantitative guidelines can lead to an increase in the quantity of children's educational programming. The Australian experience, however, has demonstrated the tendency for broadcasters to exaggerate the quantity, quality, or educational content of programming that is nominally intended for children. Australia has addressed these problems by determining in advance of a program's airing whether it meets the criteria for children's programming. Because the FCC, pursuant to the CTA, leaves the determination of whether programs meet its criteria to broadcasters subject to challenges by the public, it may be more difficult in the United States to assure that only programming meeting the definition is counted toward the guideline.

Campbell concludes with some recommendations for achieving the goals of the CTA. She argues that the Australian system in which the government preclassifies programs in advance would probably prove unconstitutional in U.S. courts. Nonetheless, she recommends that the FCC consider providing a more helpful definition of "educational," and like the Australian government, examine whether sufficient resources for program production are available and whether resulting production values are equivalent. The FCC should also, she recommends, review the efforts of licensees on an annual basis.

The final chapter in Part III is concerned with an issue of increasing importance: standard setting. David Sosa, in "AM Stereo and the 'Marketplace' Decision" (Chapter 8), challenges the widespread assumption that the FCC's decision not to set an AM stereo standard in the early 1980s prevented AM stereo from becoming economically viable. The usual argument has been that uncertainty about a standard discouraged consumers from purchasing enough sets to realize a critical mass.

Sosa presents a statistical analysis of AM stereo adoption in three major markets in which AM stereo reached substantial penetration between the late 1970s and the mid-1990s. He tests the hypothesis that audience ratings of stations that adopted stereo radio broadcasting were significantly higher than stations that did not adopt. In only one of the three markets did results suggest that stereo diffusion had any effect on audience behavior. Although Sosa's results are somewhat ambiguous, his analysis does fail, in any of the three cases, to support the conventional wisdom— that market failure was the cause. Sosa cautions against interpreting the AM stereo experience as a government failure; low consumer valuation for this change in audio quality may tell a richer story about the failure of AM stereo adoption.

IV. THE INTERNET

In Part IV, the authors of Chapters 9 and 10 deal directly with an obvious nexus of telecom policy interest: Internet telephony. The development of technology for full-duplex phone calls over the Internet brings to the fore the rapid transformation of telephony from natural monopoly to naturally competitive industry. We now see local and long distance telephony provided by traditional operators, Internet-based operators, and cable operators (e.g., in England).

In "A Taxonomy of Internet Telephony Applications" (Chapter 9), David Clark provides a much-needed characterization of Internet telephony (IPTel). IPTel is not a single physical technology, nor is it a single service offering. Ignorance about the different technologies and possible services (nearly all of them are still conjectural) causes a great deal of confusion in the trade press and current policy discussions. For example, as Clark points out, the most immediately feasible service is his Class 1, which is long distance or international calling using existing local loops, but replacing long distance or international circuits with Internet links. Although there are three reasons why this might be a cost-effective alternative, the important reason is that it can operate as a form of long-distance access charge bypass (or settlements bypass for international calls; cf. the chapters by Galbi and Scanlan). This is simply a form of regulatory arbitrage, much as we saw local-loop bypass operators perform regulatory arbitrage during the initial post-divestiture years. Fixing the regulatory inequity will leave Class 1 IPTel as a lower quality, costly-to-install alternative to traditional circuit-switched telephony.

Understanding what IPTel can do, and the role of regulatory arbitrage, is important. For example, fueled by press releases heralding massive IPTel investments by companies like Qwest and Delta3, some congressional leaders are pushing the FCC to revise the Universal Service Fund contribution rules before IPTel overwhelms traditional telephony. Clark does not address whether fixing an access charge arbitrage opportunity by changing the universal service funding base is a wise regulatory approach. Rather, he provides an extremely lucid and forward-looking characterization of IPTel necessary to address such policy questions. He illustrates this by closing with a few high-level policy implications that follow from an understanding of IPTel.

Lee McKnight and Brett Leida are Clark's colleagues at MIT; all three participate in MIT's Internet Telephony Consortium. In "Internet Telephony: Costs, Pricing, and Policy" (Chapter 10), McKnight and Leida provide a detailed economic-engineering study of advanced IPTel service (Clark's Class 3 type). This service involves end-to-end Internet communication, with the phone handset attached to the user's computer. The public switched telephony network handles no component of the voice call, although the authors assume Internet connections are obtained by dial-in service over the local switched network. New capabilities can be provided to users because communications are intermediated by powerful end-node computers. McKnight and Leida find that moderate use of computer-to-computer Internet telephony can increase the costs of an Internet service provider by as much as 50%.

Current hype about Internet telephony may make it seem surprising that IPTel can raise costs by so much. But Internet technology is based on sharing ("statistical multiplexing" is the fundamental characteristic). IPTel is not particularly well suited for sharing: it has been shown elsewhere that the ratio of bursts to average data flow is only about 2-to-1, which means that after overhead and quality control buffering, the efficiency gain from sharing cannot be much more than 1-to-1 (no sharing). Consequently, the authors estimate that holding times and call arrival rates will each increase by 20% in their main scenario. Further, customer service and billing costs tend to rise directly with new services. Thus, offering substantial IPTel service would require new capital and personnel investments. ISPs could not sustain this cost increase based on current levels of flat rate pricing. The authors believe that usage-sensitive pricing would become necessary before IPTel could commercially succeed. With these costs and the need for more revenue, it does not appear that IPTel will offer consumers large cost savings: Rather, as Clark emphasized, the main advantage of Class 3 IPTel is likely to be the value of new functionality offered to consumers through integration with an end-node computer.

In Chapter 11, "Muddy Rules for Cyberspace," Dan Burk is concerned with the evolution of intellectual property rights as digital networked distribution becomes easy and ubiquitous. Debates about intellectual property rights and digital copyright have become telecom policy issues because of the modern concern about obtaining a return for an author when digital works can be almost costlessly reproduced and distributed.

Burk points out that a critical working assumption implicit in many discussions of the copyright issue is largely incorrect: that the received wisdom embraces strong or complete property rights as the ideal. He shows that in fact rights for tangible property are often "muddy;" that is, there is ambiguity about certain claims to use property that can only be resolved through a subjective balancing test. Burk then argues that the nature of intellectual property makes muddy rules especially appropriate for certain types of use; fair use rules in copyright law are a long-standing tradition in this regard. Interestingly, Burk discusses a number of ways in which transaction costs will be higher for telecom-intermediated uses of intellectual property. This contrasts with another common preconception: that digital communications networks uniformly reduce transactions costs for commerce and exchange. In the end, Burk argues against a single "clear" rule for intellectual property in cyberspace, proposing instead that good legal rules should be as varied—and in some cases as muddy—as they are in real space.

The authors of Chapter 12 describe the interaction between engineering and sociopolitical considerations when designing a system for describing and managing privacy rights on the Internet. In "Designing a Social Protocol: Lessons Learned from the Platform for Privacy Preferences Project," Lorrie Faith Cranor and Joseph Reagle, Jr. offer an insightful case study of socio-engineering design for the Internet, as well as constructive lessons on the problem of unintended consequences. Cranor and Reagle's chapter is ostensibly about Internet concerns, but, like Burk's chapter, it in fact deals with problems that are common to policy for all communications media.

The authors have been leading participants in the collaborative project to develop a system for expressing privacy preferences on the Internet and automatically negotiating the use of personal information. For example, current technology allows Web site owners to "set a cookie" or store an identifier on a user's hard drive, which can be checked during later visits or as the user moves across the Internet to track usage and activity.[1] The proposed P3P protocol would

---

1. Most browsers can be configured to prevent cookies from being stored, but this requires some sophistication on the part of users, and makes some sites virtually unusable.

allow users to store a set of preferences specifying which types of personal information may be used for what purpose and by whom. Although the goal seems sensible, the authors clearly describe how difficult implementation can be. Along the way they show how a poorly designed communications technology can have unintended, often adverse consequences. This point is not new, of course, but Cranor and Reagle use their experience from designing P3P to constructively suggest principles for "social protocol" design that can help avoid problems. Notably, they emphasize the goal of "mechanism not policy," and then illustrate this sound principle through a series of concrete examples. One very useful lesson for policymakers is the importance of trying to distinguish between technology design decisions and policy choices.

V. COMPARATIVE STUDIES IN TELEPHONY AND SATELLITE POLICY

Learning from experience around the world has always been a strong tradition at the TPRC. We include three chapters in which the authors examine satellite policy in the Asia-Pacific region, telecommunications reform in South Africa, and differences in recent telecommunications reform between the United States and Canada.

Chapter 13, "The Paradox of Ubiquity: Communications Satellite Policies in Asia," by Heather Hudson, is critical of what she describes as politically driven satellite policies in the Asia-Pacific Region. Along with a proliferation of regional and international satellites serving the region, seven individual countries, some of them very small, now have their own satellites, four of them launched since 1993, with at least one other planned.

Hudson questions the need for these satellites, and suggests that interaction between the "national flag carrier syndrome" and liberal ITU policies for granting slots to individual countries is responsible. Hudson finds that while lip service is paid to universal service and other social objectives, in reality slots in some smaller countries have been turned over to private investors in exchange for negotiated compensation, to the neglect of social goals. Hudson concludes with recommendations on satellite policy: countries need to create incentives for investment in the terrestrial networks that will interconnect with satellite transmissions; countries need a regulatory structure in place, and agreements for interconnection with terrestrial networks; and that user needs must be accounted for more directly. These policies will at least ensure, she predicts, that the satellites will be used as effectively as possible.

Robert Horwitz's contribution, "Participatory Politics and Sectoral Reform: Telecommunications Policy in the New South Africa," (Chapter 14), tells an unusual story. We read harsh critiques about reform efforts in various countries that have stagnated or degenerated into self-interested rent seeking. In contrast, Horwitz praises the South African process of telecom reform as a politically legitimate, innovative process of consensus building among stakeholders. As a consequence, rather than being pushed aside by private interests, questions of re-distribution—notably universal service and the "general public interest"—re-mained on the front burner.

South African reform began in 1991 with the creation of a Telkom, a state-owned telecommunications monopoly, through splitting up a classic PTT. Real change began with establishment of the National Telecommunications Forum (NTF), a public participatory process modeled after other South African reform initiatives coinciding with the dismantling of apartheid following the 1994 elec-tions. Negotiations in the NTF between the main stakeholders—Telkom, labor, and business interests—resulted in draft legislation specifying a 3-to-5-year peri-od of exclusivity for Telkom over basic switched telecommunications services, to-gether with provisions for interconnection, free entry into long distance and other services, and a reformed universal service fund among other provisions. To be sure, Horwitz notes, the process had its faults. The draft bill was seriously com-promised in a ministerial review, and attention to universal service was sometimes more rhetorical than substantive. And, we should not forget that implementation of the South African reforms has yet to be accomplished. Overall, though, Horwitz describes the reform process leading to the legislation as a model process: "tech-nically viable and for the most part, politically legitimate."

The final chapter in Part V is also a study complimentary to one country's re-cent telecommunications regulatory reform. Willie Grieve and Stanford Levin, in "Telecom Competition in Canada and the United States: The Tortoise and the Hare" (Chapter 15), give high marks to the Canadian reform process leading up to a May 1997 ruling that established rules for local telephone service competi-tion. They argue that the rules will lead expeditiously to true facilities-based com-petition at the local level. The United States is likened to the hare for getting off to a much quicker start than the Canadian tortoise, but stopping before the pro-cess was complete.

Grieve and Levin believe the U.S. Telecommunications Act of 1996 has created a "shade tree" which "may actually entrench monopoly and market power in the local networks of the incumbent local carriers." The authors focus on the different approaches to unbundling and resale taken in the two largely parallel reform movements. Both the Canadian and U.S. reforms are comparable in their require-ments for interconnection of incumbent networks with those of entrants. Grieve and Levin argue, however, that the American legislation mandates excessive un-bundling, and sets unrealistically low prices for resale of unbundled network ele-ments by not requiring a sufficient contribution to an incumbent's fixed costs. The Canadian reform, on the other hand, limits unbundling and resale requirements to only the "essential facilities" of incumbents. The result, the authors claim, is that market entrants in the United States have inadequate incentives to engage in true facilities-based competition, and will simply continue repurchasing and reselling elements of the incumbent's networks without providing true competition for them. In Canada, they believe, the law encourages entrants to construct competing facilities, which many analysts agree must be the basis for true local telecommunications competition.

THE EDITORS

This book was edited while David Waterman was Associate Professor in the Department of Telecommunications at Indiana University, Bloomington, and Jeff MacKie-Mason was Associate Professor of Economics, Information and Public Policy in the Department of Economics, and in the School of Information at the University of Michigan, Ann Arbor.

I

HISTORICAL

A Novel Conference: The Origins of TPRC

Bruce M. Owen
Economists Incorporated

The 25th annual Telecommunications Policy Research Conference (TPRC) provides an opportunity to reflect on the origins and achievements of TPRC. An objective of TPRC has been to provide not merely a forum for communication policy researchers to exchange ideas, but also a channel for policy-relevant research to reach regulators and other government officials, and for the latter to convey their research needs to academics. Therefore, any discussion of the history of TPRC should be placed in the context of evolving government policy.

TPRC arose, not coincidentally, at the beginning of an extraordinary period in the history of telecommunications policy and regulation. Before the early 1970s, for example, it was unlawful for anyone but AT&T to offer public long distance service; there was no domestic satellite industry; it was unlawful for cable systems to import any but a limited number of distant signals; it was unlawful for any broadcaster or cable operator to offer pay-TV service consisting of entertainment series, sports events that had been on TV in the last 4 years, or movies less than 2 or more than 4 years old; and it was unlawful for customers to attach a "foreign" (i.e., any) device to the telephone network. More generally, it was the mainstream view that the telephone business was and ought to be a regulated monopoly, and that broadcasters were and ought to be protected from excessive competition in order to promote their ability to offer public service and especially local programming.

Further, and even more generally, the 1970s was a unique period in U.S. economic history; one in which the validity of the notion of natural monopoly and the virtues of regulation came into question. During these years academic skepticism or even cynicism about regulation, emanating especially from the Chicago School, spilled over into public debate. The result was not just communication policy reform but intercity bus, airline, trucking, and railroad deregulation; the beginnings of related reforms in the securities and financial services industries; and other deregulation initiatives. A dramatic change illustrative of the growing currency of economics took place at the Department of Justice Antitrust Division, which today employs four or five dozen Ph.D. economists. Before 1974 the Antitrust Division had no permanent staff of such economists. Similar changes occurred at the Federal Trade Commission (FTC). Many other countries have followed the U.S. intellectual lead in these matters, in some cases showing greater courage in implementing regulatory reform.

TPRC arose also during a period of extraordinary growth and change in telecommunications technology. Remote terminals of mainframe computers, geosynchronous satellites, fiber optic transmission lines, electronic switches, digital transmission and compression, the Internet, and many other advances created pressures for regulatory reform and facilitated reform.

TPRC BEGINNINGS

The institution of TPRC was neither the beginning of academic interest in communications policy nor the first time academics, lawyers, political scientists, engineers, and economists had a direct impact on communications policy. Modern academic interest in communication policy can be traced to Ronald Coase's (1959, 1962) famous property rights papers on spectrum allocation, and to such theoretical work on utility regulation as the well-known Averch and Johnson (1962) paper.

Those unfamiliar with the field will wonder what is meant by "communication" or "telecommunication" in the present context. What is meant, roughly, is those activities historically subject to the jurisdiction of the Federal Communications Commission (FCC). This usage is curious, because telephone regulation has much more in common with electricity or natural gas regulation than with broadcasting. If industry research were focused on firms with basic similarities in their products and technologies, we would have separate conferences on mass media and on public utilities. That the same research community, and even the same individual researchers, focused on the legal jurisdiction rather than the more natural economic classifications illustrates the important influences that government has on policy research.

Although important and relevant research existed, the government appeared to remain ignorant of it until the late 1960s, when Lyndon Johnson convened the President's Task Force on Telecommunications Policy, headed by Undersecretary of State Eugene V. Rostow (President's Task Force, 1968). The Task Force was established in part to hold back a rising sea of political pressure that had begun to lap at the White House gates. The pressure arose from the desire of potential entrants to arbitrage the growing gap between prices and costs or between actual and best-practice technologies, and from those incumbents who relied on government to protect economic rents. These pressures were manifest chiefly in controversies involving long distance telephone service, domestic communication satellites, and the import of distant TV signals by cable systems.

Rostow assembled a talented staff. For example, Richard A. Posner was seconded from the Justice Department and Walter Hinchman from Commerce. Leland L. Johnson came from RAND. More than 30 academic consultants were retained, including William J. Baumol, William F. Baxter, William Capron, William K. Jones, Charles J. Meyers, Monroe E. Price, and Lester D. Taylor. Government agencies sent representatives, such as Roger G. Noll from the Council of Economic Advisors. The Task Force, its consultants, and its research contractors, well aware of relevant academic research, produced a report that was cautiously progressive, suggesting for example an "open skies" policy for domestic communication satellites, and a greater role for competition in telephony. The staff and contractors also produced several innovative papers on marketable spectrum rights. Finally, the Task Force recommended establishment of an executive branch agency to formulate and coordinate telecommunication policy. More important than the specific recommendations, however, the Task Force implicitly validated the notion that there was such a thing as "telecommunications policy," that it was susceptible to analytical policy research and analysis, and that there existed a newly self-aware community of scholars interested in such research.

ESTABLISHMENT OF THE OFFICE OF TELECOMMUNICATIONS POLICY

When President Johnson did not run for reelection, his Task Force lost its constituency. Politics notwithstanding, however, the incoming Nixon administration picked up on and sought to implement many of the Task Force recommendations. Clay T. (Tom) Whitehead, a Special Assistant to the President assigned to communication matters, perhaps because he had a doctorate from MIT (in political science), pushed to implement both the satellite open skies policy and the establishment of an executive branch policy agency. The resulting Office of Telecommunication Policy (OTP) was created by Executive Order as part of the Executive Office of the President in 1970. Tom Whitehead became the first director of the agency, reporting at least in theory directly to the president.

OTP inherited the frequency management and emergency preparedness roles formerly exercised by the defunct Office of Telecommunications Management (OTM), along with many of OTM's staff. Whitehead added only a small number of new professional staff. Among them were general counsel (now Justice) Antonin Scalia, and legislative and press relations officer Brian Lamb (later to found C-SPAN). I was the first economist at OTP, initially as a Brookings Economic Policy Fellow, and later as chief economist. Other early OTP economists included Stanley M. Besen, Ronald Braeutigam, and Gary Bowman.

OTP tended to see itself, not indefensibly, as a beacon of reason adjoining an ocean of bureaucratic backwardness. Lacking significant political power (President Nixon and his senior staff did not accord much priority to telecommunications policy even before Watergate), line authority or political experience, Whitehead was reduced chiefly to issuing position papers, making speeches, and writing policy letters to the FCC chairman, which were mostly ignored. This was of course frustrating to those of us aware of the enormous gap between the implications of academic research and the actual state of communications policy in the United States.

THE 1972 CONFERENCE

Several influences led to the convening of the first telecommunications policy research conference. First, it seemed that exposing other policymakers to academic ideas might eventually make them more susceptible to OTP's positions. Second, OTP had a research budget to spend, and a conference appeared to be a sensible use of research funds. Earlier expenditures had sometimes produced embarrassing results, such as studies whose conclusions were at odds with OTP's positions. Third, because academic research appeared to be the major positive factor on OTP's side of most issues, OTP wanted to promote more of it. Giving academics a live audience of policymakers seemed likely to stimulate interest among policy scientists and their students.

Finally, to those of us with academic backgrounds, the Washington telecommunications policy community in the early 1970s was a lonely and inhospitable place. It is not an overstatement to say that ideas like "selling the spectrum" or "breaking up *the* telephone company," or even allowing competition with it, were treated with derision and contempt by responsible officials at all levels. A policy research conference would be good for morale—a booster shot for the OTP staff and the few "enlightened" analysts in other agencies.

The first TPRC was held on November 17 and 18, 1972, in the New Executive Office Building. The audience consisted of federal government employees from OTP, the FCC, and the Departments of Justice, Commerce, and Defense, among others. Papers were presented and discussed by 15 academics (13 economists and 2 lawyers). Among the most luminous academics were Ronald Coase and William Baumol. (The 1972 program is provided in Appendix A.) The research papers were published by OTP (Owen, 1972).

The topics discussed at the first conference are for the most part still on the policy agenda. There were, for example, papers on cross subsidization, financing public broadcasting, spectrum markets, and cable television regulation. There were also papers on subjects that have not been much addressed in subsequent conferences, such as democracy in the newsroom, and one paper analyzing the effect of policy research on FCC decision making. The first conference was regarded as a success by most of the participants, and there developed a consensus that it would be useful to have an annual conference.

AN ANNUAL EVENT

Although I conceived and organized the 1972 OTP conference, arguably the true beginning of TPRC was at Airlie House on April 16 through 19, 1974. (The program of the 1974 conference appears as an appendix in Owen, 1976.) Although OTP provided partial funding, this was the first independently organized meeting. The 1974 conference was organized by a group of academics (Donald A. Dunn, Stanley M. Besen, Gerald Faulhaber, Leland Johnson, and Ithiel de Sola Pool).

In later years funding came from government agencies such as OTP, the FCC, the National Telecommunications and Information Administration, and the National Science Foundation, as well as from private foundations and programs that either sponsored TPRC directly or funded research that was presented at TPRC. These institutions included the Markle Foundation, the Kettering Foundation, the Sloan Foundation, the Ford Foundation, and the Aspen Institute.

It was the practice of organizing committees in the early years to appoint their successors, with little or no overlap from year to year (organizing committee members through the 1981 conference are listed in Appendix B). Also, it was usual for the organizing committee to include representatives from those few organizations with concentrations of telecommunications policy researchers, such as the RAND Corporation, Bell Labs, and Stanford University. Each organizing committee had to manage funding as well as the program and other administrative arrangements. Because the conference had no permanent home for purposes of funding and administrative services there were frequent difficulties. By the early 1980s many established participants felt that TPRC had drifted away from its original character and goals. Accordingly, in 1987, the conference was reorganized in such a way as to separate program responsibility from fundraising and administrative concerns. Administrative matters were undertaken by a Board of Directors, whose self-perpetuating members have overlapping terms. The Board also has the duty to appoint the annual organizing committee, which has responsibility for the program and local arrangements. Since 1989 Economists Incorporated has provided administrative services to TPRC at cost; in practice this work has been organized by Dawn Higgins.

TPRC is, if not unique, certainly unusual in being a long-running event with no single individual or organization continuing in charge. Conferences like TPRC are more typically organized by learned societies. TPRC has been fortunate in having attracted such a long string of interested and capable organizing committee members. Continuing interest is no doubt also stimulated by the cataclysmic events that have shaken the communication industries since the early 1970s.

TPRC is unique in another respect: the participation of industry researchers. From the beginning, researchers from organizations such as Bell Labs have been an integral part of TPRC. Nevertheless, in the early years there was much debate, which continues, about the participation of industry "lobbyists."

INFLUENCE OF TPRC

It is difficult to say what influences TPRC has had on the development of government policy and on academic policy research because we lack a "control" world with no TPRC. Some of what we are inclined to attribute to TPRC may be due simply to the technological changes that led to revisions in telecommunications industry structure and regulation. However, in celebrating TPRC's 25th anniversary, perhaps we should not demand too much analytical rigor on this point.

One obvious and demonstrable change on the input side is the growth in the number of economists and other professionals with similar training now employed by the FCC and other agencies responsible for telecommunication. In 1970 the FCC had no more than three or four Ph.D. economists; today there are many dozens, and an even greater number are employed by regulated firms and consulting firms. Any given bureau of the FCC today is likely to employ more economists specialized in communications than there were in the nation in 1970. Further, FCC lawyers and other staffers who are not economists have adopted much of the language and many of the precepts of economics.

On the output side, changes have been revolutionary. No important FCC policy statement issues these days without explicit attention to its economic welfare effects. It is true that similar strides have been made in other areas. One is struck, for example, that at the 1997 Tokyo summit meetings on the environment, one of the United States' principal goals was the establishment of tradable emission rights. Nevertheless, communications was undoubtedly the first of the major regulatory fields to be thus reformed, and has progressed the most. TPRC facilitated this in two ways. First, by increasing academic interests in the field, it increased the supply of interested graduate students and relevant dissertations. Second, the private and government lawyers who have always been central participants in the policy process heard at TPRC a whole new set of arguments and principles that transcended the usual motifs of legal argument. Lawyers are always competing to win arguments, and TPRC supplied them with new and more effective ammunition. Further, many academic lawyers became interested in communications policy research, often as part of interdisciplinary teams.

A cynic might say that a great portion of what has changed is that the same old vested interests now feel compelled to make their public interest arguments in terms acceptable to scholars, without necessarily leading to any change in outcomes. However, such cynicism cannot explain how the preexisting industry structure was transformed into entirely new "vested" economic interests, such as

IXCs, RBOCs, CLECs, DOMSATs, and PCS licensees. Under the old regime these would all have been departments of AT&T or would not have existed at all.

TPRC's unique contribution, in the end, was the creation of what Stan Besen called an "invisible college" or virtual community of communication researchers scattered at different institutions and agencies. However characterized, TPRC promoted both academic collaboration and the delivery of relevant policy analysis to government agencies, phenomena previously unknown in the communication world.

ACKNOWLEDGMENTS

I am grateful to many of those mentioned by name herein for reviewing the manuscript and pointing out at least some of the errors.

REFERENCES

Averch, H. and L. Johnson, (1962). Behavior of the firm under regulatory constraint. *American Economics Review*, 52, 1052.

Coase, R. (1959). The Federal Communications Commission. *Journal of Law and Economics*, 2, 1.

Coase, R.(1962). The Interdepartmental Radio Advisory Committee. *Journal of Law and Economics*, 5, 17.

Owen, B. M. (Ed.). (1972). *Papers and proceedings, Conference on Communication Policy Research*. (NTIS accession number PB 218 981/9). Washington, DC: Executive Office of the President, Office of Telecommunications Policy.

Owen, B. M. (Ed.). (1976). *Telecommunications policy research: Report on the 1975 Conference proceedings*. (Aspen Institute Program on Communications and Society). Aspen, Colorado.

President's Task Force on Communications Policy. (1968). *Final report*. Established Pursuant to the President's [August 14, 1967] Message on Communications Policy (NTIS accession numbers PB 4184413-PB184424). Washington, DC: U.S. Government Printing Office.

Appendix A

OFFICE OF TELECOMMUNICATIONS POLICY
EXECUTIVE OFFICE OF THE PRESIDENT
WASHINGTON, D.C. 20504

CONFERENCE ON COMMUNICATION POLICY RESEARCH
NOVEMBER 17–18, 1972

The Conference will be held in Room 2008, New Executive Office Building, 17th and H Street, N.W., Washington, D.C.

Schedule for Formal Sessions
NOTE: Since the papers will be distributed in advance, authors will have only 15 minutes to summarize their major points. Discussants will then have 15 minutes to comment on the paper. The remainder of the time will be devoted to general discussion by the participants and members of the audience.

Friday, November 17

9:00 AM   Welcoming Remarks by Clay T. Whitehead, Director, Office of Telecommunications Policy.

9:15 AM   Edward Zajac and Gerald Faulhaber, Bell Labs: "Some Preliminary Thoughts on Subsidization."
Discussant: Thomas Moore, Michigan State.

10:15 AM   Robert Meyer, Purdue: "Public vs. Private Utilities: A Policy Choice."
Discussant: William Capron, Harvard.

11:00 AM   Ross Eckert, USC: "Spectrum Allocation and Regulatory Incentives."
Discussant: Alfred Kahn, Cornell.

11:45 AM   Lunch break.

2:00 PM   James Rosse, Stanford: "Product Quality and Regulatory Constraints."
Discussant: William Baumol, Princeton.

2:45 PM   Roger Noll, Brookings: "Decentralization of Public Television."
                 Discussant: Thomas Moore, Michigan State.

3:45 PM   Stephen Barnett, Berkeley: "Media Control, News Control, and
                 the FCC."
                 Discussant: Ronald Coase, Chicago.

Saturday, November 18

9:15 AM   Stanley Besen, Rice: "The Economics of the Cable Television
                 'Consensus.'"
                 Discussant: George Hilton, UCLA.

10:00 AM   Lionel Kestenbaum, "Issues in Common Carrier Regulation of
                 Cable Television."
                 Discussant: Merton Peck, Yale.

10:45 AM   Rolla E. Park, RAND: "The Role of Analysis in the Formation of
                 Cable Policy."
                 Discussant: Merton Peck, Yale.

11:30 AM   End of Conference.

Appendix B

Organizing Committee Members 1972–1981
The First 10 Years

1972
Bruce M. Owen

1973
No conference.

1974
Stanley M. Besen
Donald Dunn
Gerry Faulhaber
Leland Johnson
Ithiel de Sola Pool

1975
Kan Chen
Walter S. Baer
Elizabeth E. Bailey
Kas Kalba
Bruce M. Owen

1976
[Information missing]

1977
Shiela Mahony
Paul Bortz
Ronald Breautigam
Forrest Chisman
Deen Gillette
Bridger Mitchell
Weston Vivian
Debbie Mack

1978
Lawrence Day

Herbert Dordick
Aimee Dorr
Henry Goldberg
Carol Keegan
Harvey Levin
William Lucas
Rhonda Mange
Robert D. Willig

1979
[Information missing]

1980
John Clippinger
Robert E. Dansby
Charles M. Firestone
Heather Hudson
Jorge Schement
Marvin A. Sirbu
Leonard Waverman

1981
Barry Cole
Timothy Haight
Hudson N. Janisch
Wilhelmina M. Reuben-Cooke
William E. Taylor
Armando Valdez
Larry White

II

TELEPHONY

A Technico-Economic Methodology for the Analysis of Local Telephone Markets

Farid Gasmi
IDEI, Université de Toulouse I and Université de Bretagne Occidentale, Brest, France

Jean-Jacques Laffont
IDEI, Université de Toulouse I, France

William W. Sharkey
Federal Communications Commission

This chapter describes an empirical methodology for studying the regulation of local telephone markets that combines an engineering process model of costs with models from the new regulatory economics. This technico-economic methodology is illustrated through the undertaking of two analyses. First, we study in some detail the properties of optimal regulation under asymmetric information. We examine three issues: (a) the extent of natural monopoly when informational rents associated with regulation are taken into account; (b) the extent of the divergence of pricing under the optimal regulatory mechanism from optimal pricing under complete information (incentive correction); and (c) the implementation of optimal regulation through a menu of linear contracts. We find that, for fixed territory, strong economies of scale allow local exchange telecommunications to retain monopoly characteristics even when the (informational) costs of regulation are properly accounted for. Furthermore, the incentive correction term is small in magnitude, and optimal regulation can be well approximated through relatively simple linear contracts. In the second phase of our analysis, we evaluate the relative performance of various regulatory mechanisms, from both traditional and

15

modern (incentive) points of view. This analysis allows us to quantitatively assess the social value of regulatory transfers and of good cost auditing procedures, the redistributive consequences of the various forms of regulation, and the sensitivity of the relative performance of the various methods of regulation to the cost of public funds.

The importance of costing methods in regulation of network industries is by now well established. This is particularly the case in telecommunications. Historically, various approaches have been used to evaluate costs in telecommunications, including accounting and econometric methods. However, with the rapidly changing technology, forward-looking engineering models of costs have increasingly proven useful. Meanwhile, new forms of regulation have made their way into telecommunications, triggered, on the one hand, by widespread dissatisfaction with the performance of traditional forms of regulation, and on the other hand by the broad political trust in private incentives for enhancing the performance of operators. This chapter proposes a methodology that utilizes the engineering approach in a modern regulatory economics framework for analyzing some important issues facing regulation of telecommunications today.

The usefulness of this technico-economic methodology is demonstrated through two studies. First, we have used this approach to empirically investigate in detail the properties of the optimal regulation of the local telephone service and its implementation. In particular, we have explored ways to generalize the natural monopoly test in order to take account of the (informational) costs of regulation. Second, we have used this methodology to compare the performance of various regulatory schemes from both traditional and modern incentive regulation.

The plan of the chapter is as follows. The next section presents the modeling framework and characterizes each of the regulatory mechanisms. The section after that outlines the main features of the technico-economic methodology used in the empirical analysis of various methods of regulation.

The next section is devoted to the first application: the empirical analysis of optimal regulation. We suggest a way to perform a test of the natural monopoly hypothesis that takes into account the costs of regulating a monopoly under asymmetric information. These costs take the form of an informational rent that the regulator must give up to the firm. The result is that, for a fixed number of subscribers, the social benefits of reduced cost, resulting from economies of scale, may well offset these informational costs. We provide an empirical test of the incentive-pricing dichotomy property, which allows the regulator to differentiate the regulatory role of transfers from that of prices. When this property holds, pricing under incentive constraints is equivalent to (Ramsey) pricing under complete information. Given this dichotomy, we show how optimal regulation can be implemented in a simple manner. We construct a measure of firm performance of the average-cost type and show how the transfer function is decreasing and convex in this performance variable. This transfer function can then be approximated by a menu of linear cost reimbursement rules giving to efficient-technology firms higher fixed payments and lower reimbursement for cost overruns.

In the next section, we present the second application: the empirical analysis of the performance of alternative methods of regulation. After presenting some general features of the empirical data, we suggest a comparison of the various regulatory schemes along three dimensions: incentives, transfers, and cost observability. An important finding is the good performance of price cap with sharing of earnings, a method which has been widely used in telecommunications. We then analyze the welfare consequences of the regulatory mechanisms and highlight the trade-off between consumer surplus and the firm's rent. We investigate the sensitivity of our results to the price elasticity of demand and the cost of public funds. This analysis allows us to discuss the relevance of incentive regulation to less developed economies. The final section concludes and suggests some extensions of the analysis.

THE REGULATOR–REGULATED FIRM RELATIONSHIP: A UNIFYING FRAMEWORK

The new view of regulation stresses the role of asymmetric information in the analysis of the regulator–regulated firm relationship. In a framework where the regulator designs the regulatory contract, the main consequence of this information asymmetry is that the regulator must recognize the need to give up a rent to the firm (which has superior information) in order to provide it with (social welfare enhancing) incentives to minimize costs of production. This is the fundamental rent–efficiency trade-off that regulators have to face when regulating public utilities. This section outlines the basic features of a conceptual framework of this regulator–regulated firm relationship.

The production technology of the regulated firm is assumed to be described by a cost function that gives the total cost of producing various levels of output. This technology is, however, better known by the firm than the regulating authority. More specifically, the regulated firm may possess knowledge about technological parameters, such as the magnitude of fixed and marginal costs, that is unavailable to the regulator. Further, the firm may invest in some cost-reducing activity to efficiently use productive resources (e.g., labor) that the regulator cannot observe. In the former case the information problem concerns an exogenous variable (this leads to a so-called adverse selection situation), whereas in the latter case the information problem concerns an endogenous variable (this is a moral hazard situation). Hence, total cost of production is a function of these two variables as well. Except for the values of these two variables, the regulator is assumed to know this function.

The market conditions are described by the firm's technology and by a known demand function for the good produced by the firm. Maximization of social welfare, which aggregates consumers' and firm's welfare associated with the production of the good, is assumed to be the sole objective of the regulator. When the firm exerts a cost-reducing effort, the total cost of production properly includes the disutility generated by this effort. The regulatory relationship can then be modeled in a simple manner, where the regulator delegates to the firm the provision of the good to consumers. If the regulator observes ex post costs, one can adopt the accounting convention that the regulator collects the revenues from sales, reimburses the firm's production cost and gives it a compensatory payment. Because the regulator collects funds through distortionary taxation, it must account for the (shadow) cost of each dollar transferred from consumers to the firm when evaluating social welfare.

As a benchmark, we first consider an ideal world of complete information where the regulator knows perfectly the technological characteristics of the firm, and observes the cost-reducing effort. This situation is referred to with the label *CI*. In this circumstance, the regulator can merely instruct the firm to choose the output and effort levels that maximize social welfare. Society dislikes leaving rents to the firm as they must be paid for with costly public funds. Hence, the informed regulator assigns zero rent to the firm in this complete information context.

One realizes, however, that the regulator does not know with perfection the technology used by the firm. Furthermore, the regulator cannot observe how great an effort the firm is putting into improving its use of inputs. The regulator's objective is then to generate the highest social welfare possible, given these informational constraints. The optimization program is contingent on the observables (cost and output) and the beliefs that the regulator has formed about the technology of the firm. Knowing that the larger the efficiency gains that accrue to the firm, the greater the firm's incentives to produce efficiently, the regulator leaves some (informational) rent to more efficient firms. Characterizing optimally this trade-off is at the heart of optimal regulation.

Modern regulatory economics has relied on incentive theory to interpret the regulator–regulated firm relationship as a revelation mechanism, which specifies, for each *announced* technological parameter (referred to as the firm's type), a production level, a cost to realize, and an associated compensatory payment to the firm. A critical constraint, the incentive compatibility constraint, ensures that the firm is always better off announcing its true type. A further constraint ensures the feasibility of the relationship; that is that the firm agrees to participate (which is the sine qua non condition for the provision of the good at all). If aggregate costs of the regulated firm are observable ex post (e.g., through auditing) optimal regulation can be viewed as a mechanism that specifies, for each firm's type, the optimal output (or equivalently price) and effort level, and the associated compensatory payment (transfer) to the firm. This in turn determines the informational rent that should be left to the firm. This optimal regulatory mechanism with ex post cost observability has been extensively analyzed by Laffont and Tirole (1986, 1993). It is labeled *LT* in this chapter.

If costs are not ex post observable, transfers from the regulator will be independent of ex post costs, and the firm will always have the incentive to choose the optimal level of effort, namely, the one that equates marginal disutility of effort and marginal cost saving due to effort. Consequently, the expected social welfare maximizing regulator has only to determine the optimal level of output (equivalently price). Because there are fewer regulatory tools, optimal social welfare is necessarily lower. The mechanism without ex post cost observability has been analyzed by Baron and Myerson (1982). It is labeled **BM**.

Under price cap regulation, the price (equivalently output) decision is decentralized. Hence, the main objective of the regulator under this type of regulation is production efficiency, which should be a consequence of the fact that the firm is the residual claimant of any cost reductions. In order to prevent the firm from exercising its monopoly power, the regulator sets a price ceiling. The firm may, however, choose its monopoly price if this ceiling is not binding. The regulatory mechanism seeks then to determine the level of this cap that maximizes aggregate expected social welfare. This pure price-cap mechanism is labeled **PC**.

As discussed already, efficiency comes at a social cost of informational rents left to the firm. The excessive profits associated with pure price cap regulation have sometimes led regulators to append to it a profit-sharing rule. The regulatory reforms conducted in the United States during the 1980s incorporated one form or another of such sharing rules. In this case, the regulator has to simultaneously determine a price ceiling and a tax rate, the latter intended to ensure that consumers would benefit from the gains in efficiency. This regulatory mechanism is labeled **PCT**.

Under traditional cost-plus regulation, the regulator is assumed to observe ex post costs and to fully reimburse the firm for them. Clearly the firm then has no incentive to minimize its production costs and can be expected to choose its minimum level of effort. Two schemes might be considered. First, the regulator might choose not to make any additional transfers to the firm (besides reimbursing its production costs) and impose the output level that balances the firm's budget. This mechanism is labeled **C+**. Alternatively, the regulator might choose to collect revenues from sales and give the firm a net transfer that can be used to ensure that the firm's budget is balanced ex post. In this case, the regulator might advise the firm to produce the level of output that maximizes expected social welfare because having its costs reimbursed, the firm is indifferent to the level of output. This mechanism is labeled **C+T**.

A TECHNICO-ECONOMIC APPROACH FOR EMPIRICALLY ANALYZING METHODS OF REGULATION

The discussion of the previous section makes it clear that in order to evaluate the various forms of regulation, one needs to have a detailed specification, in the context of a representative local telecommunications network, of the firm's observable cost function, the market demand function, the disutility of effort function, and the distribution of the technological uncertainty parameter. In this section, we describe our methodology for choosing appropriate functional forms based on a technico-economic model.

Role of Engineering Process Models in Evaluating Telecommunications Costs

Several methods can, in principle, be used to obtain an empirical cost function. Traditionally, in a regulatory framework, detailed accounting data have been used to determine a regulated firm's embedded cost of providing service in a given service area. Although this method can yield highly detailed and disaggregated cost data, it is not suitable for defining a precise form for an underlying cost function. Econometric methods using historical accounting data have been widely used, however, to estimate telecommunications cost functions. In these studies, a flexible functional form, such as the translog, is typically assumed. By jointly estimating the cost function and a set of factor share equations using historical data, it has been possible to obtain an estimate of a historical cost function. Initially, these studies focused on aggregate time series data from large telecommunications providers and were primarily concerned with estimating the degree of scale economies in the industries.[1] Later, econometric studies using pooled time series and cross-sectional data have attempted to directly test for natural monopoly in local exchange telecommunications (see Shin & Ying, 1992).

More recently, considerable attention has been focused on the use of engineering proxy models for estimating forward-looking costs of the local exchange. One of the first such studies is due to Mitchell (1990). In this study, a heuristic model of the local exchange network was constructed, including the local loop, end-of-fice switching, tandem switching, and interoffice transport. Simple linear functional forms were assumed for individual cost components of the network, and these were calibrated either by econometric techniques where data were available, or by engineering estimates provided by Pacific Bell and GTE of California, in whose territory the study was conducted. Both capital costs (network investments multiplied by a 15% annualization factor) and operating expenses (primarily maintenance and billing costs) were considered. The end result of the study was an estimate of both the fixed costs and average incremental costs associated with each of the network components. Mitchell's study focused on estimating the incremental cost of an additional access line or an additional minute of usage of the network. These results are of potential interest in guiding investment decisions of telecommunications providers and, in a regulatory context, for setting price floors or for detecting predatory behavior.

---

1. See Christensen et al. (1983), Fuss and Waverman (1978) and Evans and Heckman (1983).

Recently, a new set of engineering process models have been introduced that seek to estimate the stand-alone cost of the local exchange network, as well as the incremental cost of a service or a network element. Two representative examples of such models are the Benchmark Cost Model (BCM) and the Hatfield Model (HM). These models are based on costing methodologies, known as *total service long run incremental cost* (TSLRIC) or *total element long run incremental cost* (TELRIC). In August 1996, the Federal Communications Commission (FCC) ruled that a TELRIC methodology should be used in setting prices charged by incumbent carriers for unbundled network elements provided to entrants.[2] Along with other proxy models, BCM and HM were evaluated by the FCC, although neither was explicitly adopted pending further study. In November 1996, a Federal-State Joint Board on Universal Service issued a Recommended Decision that concluded that proxy cost models potentially provided a reasonable technique for determining a company's forward-looking cost for the purpose of determining subsidies to high-cost areas. [3] In addition to providing point estimates of the total cost of telecommunications services as in the preceding applications, these stand-alone cost models could potentially be used to test for the presence of economies of scale and scope and for detecting cross-subsidies.

A Network Optimization Model

This section describes the particular engineering process and optimization model, the Local Exchange Cost Optimization Model (LECOM), which we have used to calibrate the cost function for local telecommunications costs. LECOM has been developed by Gabel and Kennet (1994), who used it to determine the degree of economies of scale and scope in local telephone markets and the cost-minimizing deployment of technology (e.g., analog vs. digital or fiber vs. copper). The software for LECOM combines an engineering process model, which computes the cost function for a local exchange network with a given configuration of switch locations, using an optimization algorithm which solves for the optimal number and location of switches.

LECOM allows the user to specify the size of the local exchange territory, which consists of three concentric rectangles representing a central business district, an area of mixed residential and commercial demand, and a residential district. Both the size and the population density of each region are specified by the user, as are more detailed data about calling patterns. In addition, the user is able to specify a set of technological inputs and a detailed set of input prices in order to calibrate the model to the specific characteristics of an exchange area.

---

2. "Implementation of the Local Competition Provisions in the Telecommunications Act of 1996," CC Docket No. 96-98, First Report and Order, FCC 96-325, rel. August 8, 1996.

3. CC Docket No. 96-45, FCC 96J-3, rel. November 8, 1996.

The optimization algorithm used by LECOM consists of three separate stages. The innermost stage computes loop provisioning costs, taking as given the number, type, and location of switches. For each fixed network, an engineering process model computes the cost of the facilities required to satisfy the user-specified demands. Next, LECOM optimizes over switch locations, keeping the number and type of switches constant. Finally, the algorithm optimizes over the number and type of switches. The advantage of the multistage optimization approach is that the network topology is an endogenous variable in the cost minimization problem. For each possible output vector specified, a cost-minimizing network is computed, and the resulting cost function therefore represents the long-run total cost, using currently available technology.

Calibrating Costs, Surplus, Disutility of Effort, and Regulatory Uncertainty

In our use of LECOM, we estimate the cost of a generic local exchange network consisting of 120,000 subscribers in an area of approximately 150 km^2. This represents a large local exchange area with a medium population density. In our cost function, output consists of telephone usage of the average subscriber, measured in terms of the standard North American unit of hundred call seconds (CCS).

A theoretical cost function expresses cost as a function of outputs and input prices. In addition to a large number of engineering and technical inputs, LECOM allows the user to specify multipliers for the prices of labor and capital. We use the price of capital as a proxy for the parameter of technological uncertainty, which as discussed earlier, enters the firm's cost function. We believe that this multiplier represents a plausible one-dimensional measure of technological uncertainty that has a direct impact on all of the technological variables. Similarly, we have used the multiplier for the price of labor to measure the effect of managerial effort on total cost. In this interpretation, an increase in effort leads to a reduction in the price of labor, which we interpret as an increase in the efficiency units of labor associated with a given size of workforce. The underlying assumption is that effort is primarily directed toward efficiently utilizing labor inputs. This is clearly a strong assumption, but given the dramatic reductions in labor that have accompanied changes in regulatory regimes (e.g., when state-owned firms were privatized) we believe that it is plausible. We see in the following how this specification is useful for calibrating the firm's disutility of effort function.

The next step is to specify the market demand function and the cost of public funds in order to have some measure of social surplus. Because public funds are obtained through taxation, their cost depends on the efficiency of the tax collection system. In our analysis, we use the value of 0.3 as a benchmark, suggesting that each dollar transferred to the firm costs 1.3 dollars to society.[4] As for demand, we reviewed the empirical literature on telecommunications demand (see Taylor, 1994, for an extensive survey), and specified a demand function of the exponential form. The two parameters of this demand function were found through calibration. We relied on two assumptions. First, we assumed that the elasticity of demand is equal to -0.2.[5] Second, we assumed that revenue collected from the representative customer balances the cost of serving this customer, which by using LECOM amounts to \$246.68.[6] These two assumptions yield two independent equations which we solved to obtain the explicit form of the individual demand function.

Clearly much less is known about the disutility of effort function than about market demand because this function is, by definition, unobservable. Nevertheless, it is possible to assume a functional form consistent with theory, so that this function is increasing and convex. Here, again, we have combined some institutional and theoretical information with the cost function summarizing the data generated through LECOM to calibrate a quadratic disutility function. Finally, because we do not have detailed information about the regulatory environment, we have assumed some probability distribution (the uniform distribution) to model the regulator's uncertainty about the technological parameter. The support of this distribution was assumed to be [0.5;1.5], which is the range of values for which the LECOM cost function is defined, although some sensitivity analysis to this support has been conducted.

APPLICATION I: OPTIMAL REGULATION

Our first application of the methodology described in the previous section seeks to analyze in detail the efficiency and welfare properties of the optimal regulatory mechanism LT.[7] Three issues are examined in turn. First we examine the issue of natural monopoly in local telephone when informational rents associated with regulation are explicitly accounted for. Then we investigate the extent of incentive correction, which expresses the divergence of pricing under imperfect information optimal regulation (LT) from pricing under complete information regulation (CI). The analysis of optimal regulation concludes by proposing a way to implement the scheme.

---

4. The value of 0.3 for the cost of public funds has been shown to be reasonable for developed countries with efficient taxation systems.

5. Taylor (1994) reports a demand elasticity in the range [0.17;0.38] in absolute value. In our base case we use the value of 0.2 at an output $q=4CCS$. We have, however, performed some sensitivity analysis of the results to the value of elasticity (see below).

6. Given changing technologies and liberalized regulatory regimes, we believe that this assumption is appropriate for future regulation of local exchange companies.

7. See Gasmi et al (1997a) for details.

The Natural Monopoly Hypothesis Under Asymmetric Information

We first consider the implication of our analysis for assessing the relative merits of policies that promote entry into the local exchange market versus the maintenance of a regulated monopoly. Early empirical investigations of telecommunications technology have tested for economies of scale, and more recent work has attempted to test directly whether or not the industry, or segments of the industry, is a natural monopoly. For the case of the local exchange network, Shin and Ying (1992) proposed a direct test of subadditivity of the cost function that led them to conclude "that the cost function is definitely not subadditive" (p. 181). They further conclude that their results "also support permitting entry into local exchange markets" (p. 181).

Because industry output need not remain constant after entry occurs, a broader test of the benefits or costs of entry would account for changes in consumer welfare associated with entry. An additional benefit of entry is the reduction or elimination of informational rents to a monopoly firm. In this section we describe an analysis that incorporates each of these factors in a comparison of social welfare under optimally (LT) regulated monopoly with social welfare under both regulated and unregulated duopoly.

In our model of regulated symmetric duopoly, we assume that two firms with the same technological parameter are allowed to serve the local exchange market. We further assume that output is shared between the firms so as to minimize total industry cost, which, given our cost function, results in equal market shares. The regulator is able to offer each firm a fixed price contract in which the transfer to each firm is based on a relative performance measure. In this framework the complete information outcome can be attained because yardstick competition reveals the uncertain technology parameter and the regulator infers effort by observing cost and output from the cost function. The regulator then determines output and effort that maximize social welfare.

In our model of unregulated symmetric duopoly, we assume that the firms compete in quantities. In the absence of regulation, each firm can be expected to choose the effort level that minimizes total cost. Hence, each firm determines optimal quantity and effort by setting marginal revenue equal to marginal cost and marginal disutility equal to marginal cost saving, respectively.

We have calculated (total industry) output and firms' effort under the three market structures discussed earlier and for different types of firms. To allow for a completely symmetric comparison between monopoly and duopoly, we also calculated the value of these variables for an unregulated monopoly. The values of these variables allowed us to evaluate social welfare.

The results found indicate that, for the local exchange network considered in our study, optimal regulation of the monopoly outperforms both the regulated and unregulated (Cournot) duopoly. Duopolistic competition (with and without regulatory oversight) leads to higher industry output than unregulated monopoly and, hence, to higher social welfare despite the lower effort exerted by the firms. Regulating duopolistic competition improves the allocation of effort in the industry and strengthens the superiority of duopoly over unregulated monopoly.

Regulated duopoly necessarily dominates unregulated duopoly in our framework because we have given the regulator appropriate instruments to duplicate the unregulated duopoly outcome. Industry output and social welfare are highest under optimal regulation of the monopoly for all types of firms. Because of the incentive constraint imposed under optimal regulation, effort levels under this regulatory regime are higher than under regulated duopoly for efficient-technology firms and lower for inefficient-technology firms. From an ex ante standpoint (which is the most relevant for public policy purposes) the monopolistic structure dominates the duopolistic market structure. These results suggest that the benefit from economies of scale in the monopoly structure dominates the cost of the informational rent it is associated with.

As we were able to evaluate social welfare only after providing a value for the elasticity used to calibrate demand, a distribution for the technological parameter and a value for the cost of public funds, we ask the question of whether the outcome of this comparison of market structures is sensitive to the specific assumptions made. We duplicated this natural monopoly test for two higher values of demand elasticity than in the base case above (-0.4 and -0.6). Although the relative welfare rankings were slightly different, optimally regulating the monopoly remained unambiguously a dominant policy.[8]

Using an alternative distribution for the technological parameter, namely a truncated normal (instead of a uniform) distribution, as well as alternative values for the cost of public funds, namely 0, 1, and 10, did not affect the relative welfare rankings. The normal distribution had the effect of increasing expected output and effort and, hence, social welfare for the optimal regulatory mechanism, suggesting that the regulator had faced a "better" information structure.[9] As to the effect of the cost of public funds, when it is equal to zero, optimal regulation coincides with the best, and increasing it decreases both output and effort (and social welfare) in all of the three market structures, although the relative ranking remains the same.

---

8. In fact, for very large elasticities and for certain types of firms (efficient-technology firms), unregulated monopoly comes to dominate (in terms of social welfare) both unregulated and regulated duopoly. This is no longer the case if *expected* social welfare is the criterion used in the comparisons.

9. The issue of the value of information to the regulator certainly deserves a more careful investigation.

Incentives and Pricing Under Optimal Regulation

As discussed previously, optimal regulation under asymmetric information requires the regulator to give up a rent to the firm. A regulator with complete information would set production at a socially efficient level by equating the marginal social value of production with the marginal social cost of production. In addition, a regulator with complete information would ensure that the effort levels are set at their cost-minimizing level. Under asymmetric information, however, the regulator's desire to minimize the transfer to the firm is in conflict with the desire to have the firm minimize its costs. This conflict may give rise to an incentive distortion (i.e., a divergence of pricing from Ramsey pricing).

The magnitude of this distortion, difficult to evaluate analytically, can be estimated by using the empirical functions produced by using our methodology. The empirical results reveal that the incentive correction term is very small, never exceeding 0.63% of social marginal cost in absolute value.[10] Whenever the incentive correction is identically zero, in which case the "incentive-pricing" dichotomy is said to hold, a regulator uses the available two policy instruments, the transfer and the regulated price, for separate ends. Prices can be set to maximize social welfare and the transfer to the firm is used to optimally determine the trade-off between leaving rent to the firm and giving the firm an incentive to minimize cost.

A necessary and sufficient condition for the incentive-pricing dichotomy to hold is that the effects of the efficiency (technological) parameter and of effort on cost can be aggregated and separated from that of output. From an empirical standpoint, one can test for this aggregation property by testing for the significance, in the cost equation, of the coefficients associated with the cross-effects between the output and both the technological parameter and effort. We tested econometrically the significance of these cross-terms and found that they are statistically insignificantly different from zero at a 5% critical level. Moreover, an examination of the separate contribution of each of these variables to the explained sum of squares of the dependent variable (cost) confirms that the explanatory power of these variables is small indeed. Hence, our empirical results suggest that a LECOM-type technology satisfies the incentive–pricing dichotomy.

Implementation of Optimal Regulation

Given that we have available an actual cost function of a typical local exchange network, it is worthwhile to investigate further the issue of implementation of an optimal regulation program. As was demonstrated in the previous section, our empirical cost function satisfies the incentive–pricing dichotomy. Hence, optimal prices can be set independently from incentive issues (according to a Ramsey

---

10. The incentive correction term is nil for the most efficient firm.

rule) and optimal effort levels are determined by the transfer from the regulator to the firm. In some cases the optimal transfer can be implemented by a menu of linear cost-sharing rules.

An attractive feature of linear cost-sharing rules, besides the fact that they are relatively easy to implement, is that they are robust. Under risk neutrality of the agents, these rules are unaffected by the presence of an additive accounting or forecasting error in the cost function, and changes in the distribution of this noise do not affect their optimality. A sufficient condition for linear implementation of the regulatory transfer function is that the cost function can be expressed as the product of two subcost functions. This condition is clearly stronger than the separability property that we found to hold in the previous section. An econometric test on our empirical translog cost function showed that this sufficient condition is fulfilled.

Given this empirical result, we propose a way of implementing the optimal regulatory mechanism through a menu of cost reimbursement rules. The estimated translog function allows us to construct an explicit transfer function. Optimal transfer is made dependent upon a measure of performance of the firm, which can be thought of as a generalized average cost. This optimal transfer function is decreasing and convex in the measure of performance. Hence, it can be approximated arbitrarily closely by a menu of linear cost reimbursement rules that specifies a fixed payment to the firm and a share of cost overruns that the firm will have to support. Convexity of the transfer function ensures that firms operating with a more efficient technology receive a higher fixed payment and a lower reimbursement of the cost overruns.

APPLICATION II: COMPARING THE PERFORMANCE OF ALTERNATIVE REGULATORY REGIMES

In our second application of the technico-economic methodology, we compare the relative performance of the various methods of regulation discussed earlier.[11] The (LECOM) cost model as summarized by our estimated translog cost function and the calibrated subsidiary cost functions allow us to solve explicitly for the various regulatory schemes. Some important features of the data obtained are in order.

Under the LT mechanism the resulting output levels correspond to monthly charges that range from $8.07 to $8.63. Under C+ and C+T there is greater variation in prices, which range from a low of $5.02 under C+ for the most efficient firm to a high of $8.88 under C+T for the least efficient firm. The resolution of the pure price cap regulatory program yields a price cap equal to $6.71 which is binding (i.e., smaller than the monopoly price) for all types of firms, whereas the price cap with sharing of earnings regulatory program leads to a tax rate applied to the firm of 19.85% and a cap of $6.90, which is also binding for all types of firms. As ex-

---

11. See Gasmi et al (1997b) for more details.

pected, both types of cost-plus schemes (C+ and C+T) require higher production from more efficient firms, and this leads to higher consumer welfare.

A striking feature of the results is the high output level induced by the regulatory schemes PC, PCT, and C+ relative to that induced by all of the other schemes. This may be explained as follows. For the regulatory mechanisms that allow for transfers from consumers to *general revenue*, namely CI, LT, BM, and C+T, in low elasticity markets (elasticity equal to –0.2) society finds it worthwhile to subsidize the general activity through local exchange revenues. This is achieved by restraining quantity (and thus increasing prices). In our base case we find that local exchange revenues consistently exceed total costs under these mechanisms. In contrast, in high-elasticity markets (elasticity equal to –1.0), the subsidy goes from the general revenue to local exchange business. Indeed, we find that in the high-elasticity case, output is lower for the mechanisms PC, PCT, and C+ than for CI, LT, BM, and C+T and costs exceed revenues under these mechanisms.

Under the complete information benchmark (CI), the more efficient firms produce higher output and hence consumers receive higher welfare. Effort, which equates marginal disutility and marginal cost saving, decreases with efficiency for the less efficient group of firms and increases with efficiency for the more efficient firms. All of the endogenous variables are monotonically decreasing in the technological parameter for the optimal regulatory mechanism with ex post cost observation (LT). More efficient firms are induced to produce higher output and exert higher effort allowing them to trade-off higher consumer welfare for higher informational rents. The optimal regulatory mechanism without ex post cost observation (BM) also has the more efficient firms produce more. As in the case of complete information, effort is U-shaped.

There is no distortion in effort (relative to CI) for the most efficient firm under either LT or BM regulation. Because cost is unobservable to the regulator under the BM scheme, and is therefore fully borne by the firm, effort is socially optimal for every type of firm, conditional on output in this scheme. Indeed, we obtain that that effort is higher under the BM mechanism than under LT. Correspondingly, the informational rent to be given up to the firm is higher under BM.[12] Higher rents are generally associated with lower consumer welfare. We also find that despite the higher total cost under BM (due to higher disutility of effort), the higher rents lead to a higher rate of return than under LT. As we see later, however, the downward distortion in effort under LT is more than offset, from a social welfare point of view, by the higher informational rent extracted from the firm. These results are not surprising as the option of ignoring cost observations (BM) is also available to the regulator under the (LT) scheme.

Under PCT, part of the financial profits of the firm are returned to society.[13] Consequently, the firm exerts lower levels of effort than under pure price cap PC.

---

12. However, no rent is left for the least efficient firm.

The sharing mechanism leads to lower rents and rate of return and consumers' welfare is higher.

To conclude this presentation of the main features of the data produced by the resolution of the various regulatory schemes, we assess the relative importance of transfers from the regulator to the firm under the optimal regulatory mechanism LT. These transfers could also be viewed as representing the fixed part of a two-part tariff in situations where transfers are not explicitly allowed.[14] The fixed part would then represent a portion of total revenue that ranges from 25% for low-cost firms to 4% for high-cost firms.

The alternative regulatory schemes described are characterized by various kinds of institutional assumptions that constrain their performance relative to the complete information regulation (CI). Next we describe a framework for comparing these schemes. We then analyze their redistribute consequences and report the results of some sensitivity analysis.

A Comparison Framework

The actual regulatory schemes discussed earlier are subject to constraints of three types. Mechanisms differ in the (ex post) observability of costs, in the feasibility of lump sum transfers to the firm, and in the degree of bounded rationality of the regulator (in addressing incentive issues). Therefore, conceptually, the mechanisms can be visualized as a pair of two-dimensional transfer-observability diagrams (Fig. 2.1). Along the transfer axis there are three possibilities: no transfer, one-way transfer, and two-way transfer. Along the observability axis there is (ex post) cost observability and effort observability (which applies only to the complete information benchmark).

The upper diagram in Fig. 2.1 shows the schemes falling into incentive regulation. At the origin is the (pure) price cap mechanism (PC) with no observability and no transfer. The BM scheme is two steps up along the transfer axis to reflect the possibility of two-way transfer. PCT is obtained by moving, from the origin, one step to the right because of cost observability and one step up because of one-way transfers. The LT mechanism is obtained by moving one step to the right from BM along the observability axis, to account for ex post cost observability. One step further to the right along this same axis is located the CI scheme in which effort is also observable. The lower diagram represents the schemes falling into tra-

---

13. Since the regulator cannot observe effort levels, financial rather than actual profits are used as the basis for profit sharing.

14. Note, however, that this would require a different interpretation of the deadweight loss associated with transfers, which here would have to account for the possibility of disconnection of low-income consumers. See Gasmi et al. (1994) for a discussion of the assumption of allowance of transfers.

ditional regulation. For both C+ and C+T, cost observability is assumed and these schemes are differentiated only by whether or not transfers are allowed.

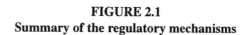

FIGURE 2.1
Summary of the regulatory mechanisms

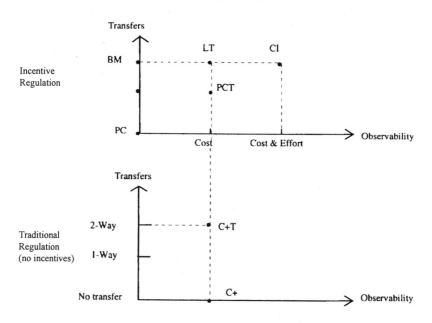

In the remainder of this section, we compare the various mechanisms by assessing their relative performance, measured in terms of expected social welfare. Because the range of parameters used in the simulations somewhat limits the variability of the results, it is instructive to draw comparisons in relative rather than absolute terms. Hence, we have chosen to analyze the performance of the various regulatory mechanisms relative to that of the regulation that would be implemented in a complete information world (CI). To assess the (social) gain associated with the use of the LT mechanism relative to the C+ mechanism, we first calculate the loss in (expected) social welfare, with respect to the complete information social welfare, associated with cost plus regulation. Then we find that optimal regulation with ex post cost observability (LT) allows the regulator to recover 72.91% of this loss in welfare.

We can use this technique to compare each of the possible forms of incentive regulation with each of the traditional forms of regulation. We found that in each of the eight possible pairwise comparisons, more than 60% of the welfare loss associated with traditional regulation can be recovered by switching to incentive regulation of one form or another. The need to create incentives for cost minimization has been one of the main driving forces behind recent regulatory reforms in developed countries. Our empirical results both explain and justify the popularity of these reforms.

Using the same techniques, we can also measure the gains associated with observability of cost and with the regulator's ability to use transfers. A comparison of the LT and BM mechanisms can be interpreted as an indication of the value of good cost auditing procedures, assuming that accounting manipulations can be easily detected.[15] Using our figures, we have found that the use of ex post accounting data allows the regulator to recover 11.19% of the loss associated with the BM regulatory mechanism. A comparison of the PC mechanism with either LT or BM shows the consequences (resulting in 23.7% and 14.1% welfare loss, respectively) of constraining the regulator to not use direct transfers to the firm, a constraint that is often institutionally imposed for political reasons or by fear of capture. A similar comparison can be made between the C+ and C+T mechanisms (resulting in 10.9% welfare loss). Finally, a comparison of PCT and PC demonstrates that the use of one-way transfers and cost observability simultaneously allows the regulator to recover 10.17% of the loss associated with simple price cap regulation.

Welfare Economics of Regulation

We now examine the redistributive consequences of the various forms of incentive regulation in more detail. Price cap regulation (under the name of RPI-X) was introduced in the early 1980s as an alternative to traditional rate-of-return regulation. The innovative feature of this type of regulation is that pricing decisions are completely decentralized. Consequently, transfers from the regulator to the firm are not used. The main objective of establishing price ceilings is to simultaneously restrain monopoly power and to give incentives to the firm for cost minimization. Indeed, because under price cap the firm is the residual claimant of any cost reductions, much efficiency is achieved with PC. Indeed, our empirical results show that higher effort and quantity levels are induced under PC than under both BM and LT. The pitfall associated with PC, however, is that higher rents are realized by the firm, leading to lower consumer welfare and social welfare. This shows that efficiency achieved through decentralization comes at a social cost.

Given the high performance of the PC mechanism in terms of incentives for efficiency, profit-sharing rules have been introduced to improve its working in terms of rent extraction. The direct comparison of the PC and PCT mechanisms (see the preceding discussion), however, suggests that the gains associated with the use of transfers to the firm and cost observability (equal to 10.17% of the difference be-

---

15. If the project is small relative to the size of the firm, this assumption may fail to hold, and cost observation may prove less useful than our computations suggest.

tween aggregate welfare under CI and PC) are relatively modest, particularly when compared to the substantial gains associated with adopting any form of incentive regulation over traditional regulation. How then are we to explain the wide use of sharing rules that have been adopted whenever price cap forms of regulation are adopted? We argue that the answer may lie in a deeper analysis of the distributional consequences of imposing incentive regulation.

Our figures convey the message that, although consumers benefit most from cost-plus regulation, expected social welfare is lowest under this traditional form of regulation. In other words, if higher social welfare is the main objective, then to achieve this goal through higher production efficiency, society needs to sacrifice consumer welfare in favor of rents. This fundamental trade-off is shown in Fig. 2.2, which reports the (ex ante) distribution of social welfare between consumers (surplus) and the firm (rent) under each of the regulatory regimes. This figure may provide some empirical substance to a positive theory of the resistance of various interest groups to certain forms of regulation, which, in particular, would predict that incentive regulation is promoted by firms.

FIGURE 2.2
Rent–surplus trade-off.

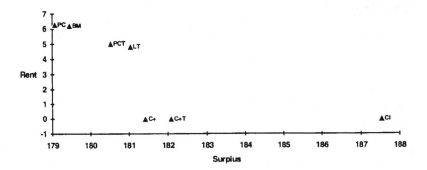

We now analyze more formally these issues. In all of the regulatory mechanisms considered earlier, the objective of the regulator is the maximization of a Benthamite social welfare function (i.e., the unweighted sum of consumers' and the firm's welfare). A generalization of this objective would consist of assigning a weight not necessarily equal to one to the firm's utility in the social welfare function.[16] This generic welfare measure has been used to generate an *Incomplete (second-best) Information Pareto Frontier*, which is exhibited in Fig. 2.3 below.

---

16. See Baron (1989) for a good discussion of the rationale behind this specification.

Because the mechanisms BM, PC, PCT, C+, and C+T impose additional constraints on the regulator, the corresponding surplus–rent pairs lie inside the second
best frontier. Indeed, one can readily show that, for any of the social welfare allocation points corresponding to one of the regulatory mechanisms discussed earlier,
there exists a firm's utility weight that leads to a welfare allocation on the second
best frontier that (weakly) Pareto dominates it. Accordingly, the locus of these
points constitutes a third best Pareto frontier.

A worthwhile observation is that the price cap regulatory mechanism with sharing of earnings (PCT) Pareto dominates a convex combination of the two extreme
points of the second-best frontier (labeled LT^{cc} in Fig. 2.3). One can interpret (see
Laffont, 1996) the allocation LT^{cc} as the result of a political process controlled
with some given probabilities by the lobbies of rent owners and consumers who
would then seek to achieve their most favored LT mechanism. Hence, constitu-
tionally imposing PCT leads to an allocation that Pareto dominates the outcome
of such a political process in an ex ante sense.

FIGURE 2.3
Second- and third-best Pareto frontiers.

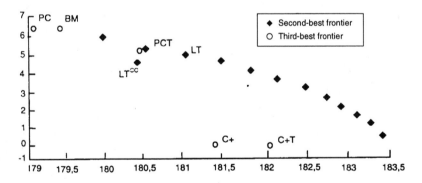

Sensitivity Analysis to Demand Conditions and Cost of Public Funds

Because it reflects the social cost associated with transfers from consumers to
the firm, the shadow cost of public funds plays a crucial role in the analysis of
welfare. Although a study of the factors that determine this parameter is beyond
the scope of our current analysis, we expect its value to be higher for economies
with less efficient taxation systems, such as developing countries. In the next
exercise, we seek to investigate empirically how this parameter affects regulatory outcomes for both low-elasticity and high-elasticity markets. We computed expected social welfare for each of the seven regulatory mechanisms
discussed earlier for various values of the cost of public funds, assuming de-

mand elasticities of -0.2 and -1.0. Various pieces of information have been revealed by these calculations.

Clearly, the cost of public funds has an impact on the performance of each of the regulatory mechanisms with transfers. In fact, for small values of this parameter, social welfare is decreasing. However, for larger values, it becomes more and more attractive to restrict quantity (the small elasticity permitting) in order to extract revenue from the local exchange business to subsidize the general budget. This U-shape of social welfare as a function of the cost of public funds is maintained for high elasticity markets. However, because it is more costly, in terms of foregone consumer surplus, to extract revenues from local exchange customers, social welfare declines over a larger range of values of the cost of public funds.

What are the implications of the shape of the social welfare function for the relative value of transfers as a regulatory tool in developing countries? A close examination of the data suggests that, in low-elasticity markets, transfers are relatively more attractive for less developed economies, whereas in high-elasticity markets, the situation is reversed, so that transfers are relatively less desirable for the less developed economies. If the allowance of transfers is already acquired, we find that, in low-elasticity markets, it might actually be the case that (in less developed economies) a careful use of these transfers to balance the firm's budget in a cost plus regime (C+T) outperforms an incentive-based regime (PCT). This is not, however, the case for high-elasticity markets. Finally, we find that good auditing procedures (to achieve observability of costs) are desirable for less developed economies in both low- and high-elasticity markets.

CONCLUSION

In view of both the state of regulatory economics and the recent structural changes experienced by industries traditionally considered as natural monopolies, this chapter has proposed a technico-economic methodology, the forward-looking aspect of which might prove useful. The recent debate on the financing of universal service in telecommunications in both the United States and the United Kingdom (see Oftel, 1997; and Telecommunications Act, 1996) has clearly demonstrated the need to adopt a forward-looking approach in evaluating the network costs associated with the provision of service. Engineering optimization models of the type used in this study might constitute a good basis for evaluating these costs. We have also demonstrated how this type of engineering model can be combined with modern regulatory economics analytical tools to both illustrate important theoretical results and produce some new empirical insights.

The methodology described in this chapter should allow for an enlargement of the analysis of the issues facing telecommunications today and in the foreseeable future. In a sector where technological developments are rapid, the location of bottlenecks will be expected to change. Moreover, the proliferation of networks allowed by the liberalization of the supply of telecommunication services and the existence of substantial network externalities will necessitate the interconnection of the networks. The regulatory implications of these technological changes and of this necessary interconnection of networks are two issues that we intend to explore next by using our methodology.

ACKNOWLEDGMENTS

Parts of this material were also presented at the North American Summer Meeting of The Econometric Society, Caltech, Pasadena, CA, June 1997, the Economics Department of the University of New Mexico, Albuquerque, NM, August 1997, as well as the 25th Annual TPRC. We thank participants for comments. Financial support of France Télécom-CNET, Issy-les-Moulineaux, is gratefully acknowledged. The views expressed in this chapter are not necessarily those of France Télécom or the FCC.

REFERENCES

Baron, D. (1989). Design of regulatory mechanisms and institutions. In R. Schmalensee & R. Willig (Eds.), *The handbook of industrial organization* (Vol. 2, pp. 1349–1447). Amsterdam: North Holland.

Baron, D., & Myerson, R. (1982). Regulating a monopolist with unknown costs. *Econometrica, 50*,911–930.

Christensen, L. R., Cummings, D., & Schoech, P. E. (1983). Econometric estimation of scale economies in telecommunications. In L. Courville, A. de Fontenay, & R. Dobell (Eds.), *Economic analysis of telecommunications: theory and applications* (pp. 27–53). Amsterdam: North Holland.

Evans, D. S., & Heckman, J.J. (1983). Multiproduct cost function estimates and natural monopoly tests for the Bell System. In D.S. Evans (Ed.), *Breaking up Bell* (pp. 253–282). Amsterdam: North Holland.

Fuss, M., & Waverman, L. (1978). Multi-product, multi-input cost functions for a regulated utility: the case of telecommunications in Canada (Working Paper No. 7810). Toronto: Institute for Policy Analysis, University of Toronto.

Gabel, D., & Kennet, D. (1994). Economies of scope in the local exchange market. *Journal of Regulatory Economics, 6*,381–398.

Gasmi, F., Ivaldi, M., & Laffont, J. J. (1994) Rent extraction and incentives for efficiency in recent regulatory proposals. *Journal of Regulatory Economics, 6*,151–176.

Gasmi, F., Laffont, J. J., & Sharkey, W. W. (1997a). Empirical evaluation of regulatory regimes in local telecommunicatons markets (IDEI working paper). Toulouse, France: Institute D'Economie Industrielle (Institute of Industrial Economics).

Gasmi, F., Laffont, J. J., & Sharkey, W. W. (1997b). Incentive regulation and the cost structure of the local telephone exchange network. *Journal of Regulatory Economics, 12,* 5–25.

Laffont, J. J. (1996). Industrial policy and politics. *International Journal of Industrial Organization, 14,* 1–27.

Laffont, J. J., & Tirole, J. (1986). Using cost information to regulate firms. *Journal of Political Economy, 94,* 1–49.

Laffont, J. J. & Tirole, J. (1993). *A theory of incentives in procurement and regulation.* Cambridge, MA: MIT Press.

Mitchell, B. M. (1990). *Incremental costs of telephone access and local use* (Rand Report R-3909—ICTF). Santa Monica, CA: Rand Corporation.

OFTEL (1997). Universal telecommunications services (Consultative Document). United Kingdom: Office of Telecommunications.

Shin, R. T., & Ying, J. S. (1992). Unnatural monopolies in local telephone. *Rand Journal of Economics,* 23(2), 171–183.

Taylor, L. D. (1994). *Telecommunications demand in theory and practice.* Boston: Kluwer.

Telecommunications Act of 1996, Pub. L. No. 104-104, 110 Stat. 56, codified at 47 U.S.C., 151 et. seq.

Firm Interaction and the Expected Price for Access

Judith A. Molka-Danielsen
Molde College

Martin B. H. Weiss
University of Pittsburgh

One of the important goals of the Telecommunications Act of 1996 is competition in the local exchange. As this competition is introduced, traditional subsidies, explicit and implict, will most likely disappear. For example, geographic averaging created an arbitrage opportunity in urban areas, one that has been taken advantage of by entrants since the early 1990s. Competition brings pricing flexibility, so that one could expect implicit subsidies, like geographic averaging, to disappear. For policymakers, an important question is, "What happens to universal service as pricing flexibility is introduced?" A large impact on universal service would imply the need for a system of explicit subsidies to meet public policy goals.

In this chapter, we consider what impact cost-based pricing (without geographic averaging) in the absence of other subsidies might have on universal service. To do this, we estimate what prices consumers might pay (and the associated penetration rates) for local access if two firms vie for their business assuming no universal service subsidy. Given the parameters of the study, the expected penetration rates for households within the study areas would be substantially lower than the current U.S. average household penetration rate, which suggests that liberalization and deregulation of local access pricing requires a corresponding need for access price subsidies if current penetration rates are to be maintained.

Put formally in this chapter, we conduct an evaluation of expected price for access to local telephone service when there is duopoly competition.[1] Using Green and Porter's (1984) model, we compute the quantities the two firms will choose to produce to maximize their present and future expected value given their strategies. We apply the resultant prices to a model for access demand that was developed by Perl (1983, 1984) to estimate penetration rates.

Many technologies exist to implement facilities-based networks that might compete with incumbents, such as CATV companies, fixed wireless, and so forth, which suggests that the cost base of the incumbent and that of an entrant would be different. We chose a CATV provider, for which Reed (1993) developed a cost model. For the incumbent's cost, we use our own study of investment costs to compute the total capital costs of access and, using U.S. Census data, compute an average network cost per subscriber for 14 specific service areas.

In addition to high network investment costs for new entrants, new entrants must match the large service portfolios of the incumbent local exchange carriers (LECs), which will increase the entry costs for local access. Because of these high costs, we assume the number of facilities-based competitors in most geographic markets will be few, so we consider the duopoly assumption to be reasonable for many markets.

We then simulate the behavior of two competing providers using an interactive game model where parameters were selected to match a current U.S. local access telephone market. We compute the expected price for access for a household and compute the future expected value to the providers.

REPLACEMENT COST DATA

For this study, we developed our own cost data so that we could explicitly link these data to data from the U.S. Census. To provide a comprehensive simulated service area, we collected data from two firms and combined them to represent one simulated service area. Firm α provided data on one of its urban service regions containing 78,371 households distributed over 10 service areas, each of which supported a central office node (which may be an end office switch or a remote switching unit). Firm β provided data on one of its rural service regions containing 6,490 households distributed over 4 service areas, each of which also supported a central office node.[2]

The two firms gave us facility investment data in the form of 1993 U.S. dollar replacement costs. Other studies that examine incremental access and usage costs have distributed investment costs over the average life of the investment (Hatfield Associates, 1994, Mitchell, 1990). They present the data in a cost per line per month format. Because we did not have usage costs, we limited our study to capital costs only. When we used the estimated cost function in the interactive model (described later), the network costs were levelized over a period of 5 years. Thus,

---

1. Special thanks to Stéphane Pallage, of CREFE and University of Quebec at Montreal, Department of Ecomonics, for his critical review of the described game.

TABLE 3.1a
Firm α: Urban Area Details of Costs

Co	Distribution Link Cost	Drop Line Cost	Remote to Host Cost	Switch & Soft Cost	Total Net Cost	Number Lines	Number Households
A	1932718	3516500	0	5096443	10545661	0	26419
B	1020992	3622000	0	4324899	8967891	28976	15585
C	616415	3950000	37511	8011259	12615185	31600	16024
D	1268762	2192000	14206	5376567	8851535	17536	9985
E	533074	248468	28412	1688932	2486168	1866	1372
F	105497	293750	20904	2024818	2444969	2350	346
G	394879	528000	74463	3694285	4691627	4223	3539
H	1144764	798750	87079	5878378	7908971	6390	1740
I	175991	638750	36562	3540422	4391725	5110	1713
J	147743	239750	40720	2072316	2500529	1917	1648

TABLE 3.1b
Firm β: Rural Area Details of Costs

Co	Distribution Link Cost	Drop Line Cost	Remote to Host Cost	Switch & Soft Cost	Total Net Cost	Number Lines	Number Households
K	1891368	425125	NA	1587500	3904011	3401	2267
L	1350990	307500	NA	1362500	3020990	2460	1640
M	1188871	271000	NA	845000	2304871	2168	1445
N	972713	213375	NA	755000	1941088	1707	1138

we present the data in the form of total network replacement costs in each service area[3] (Molka-Danielsen & Weiss, 1996).

The network elements for basic telephony access that were included in this analysis are:

2. For both firms, the central office node can be either a central office switch or a remote access node. The remote nodes can provide single-line and single-party access to the first point of switching in the local exchange network. This definition of a service area is similar to the definition that was used by Bellcore for defining Carrier Service Areas (CSA). In the Bellcore definition, the CSA can be served by Digital Loop Carriers (DLC) which use T1 remote terminals to connect subscriber loops to central office equipment (Bellcore, 1994). In our study we also include the Remote-to-Host cost, but the technology can be other than DLC systems.

3. The analysis is presented so that the identity and service locations of Firm α and Firm β remain anonymous at their request.

- **Distribution Link Cost**, which is the material cost of the cable from the host or remote node to the pedestal and the labor cost of installing the cable.
- **Host-to-Remote Cost**, which includes cable, termination, and trunk facilities that connect hosts with simpler remote node.
- **Switch and Software Cost**, which includes the hardware cost of the host or remote node located at a central office (CO) site and distributes the software costs according to the number of lines served by this site. The remote nodes that are connected to the host share the remaining software costs. The software costs are distributed from the main switches to the remote nodes because the remote nodes share some of the functionality of the main hosts.
- **Drop Line Cost,** which is the total cost for drop lines from the pedestal to the households. It is based on the number of pairs of lines multiplied by a cost per pair.

The details regarding how the cost data were collected are reported in another paper by Molka-Danielsen and Weiss (1996)[4]. The Total Network Replacement Cost was computed by summing the Distributed Link Cost, Host-to-Remote Cost, Switch and Software Cost, and Drop Line Cost for each service area. These data are presented in Table 3.1a[5] and Table 3.1b.

SELECTION OF A MARKET COST FUNCTION

To select a market cost function, we looked for a relation between the total network replacement costs and the number of lines in the service area. A regression line fitting was applied to the replacement cost data from our 14 study service areas. We found that the best fit to the data was obtained using a power curve; the estimated parameters of the power were $B_0 = 36783$ and $B_1 = .5597$,[6] which fit

---

4. The Remote-to-Host Cost includes only the trunk costs between the Remote Switch Concentrator (RSC) or a Remote Line Concentrator (RLC) and an end office switch. For all of the service areas that use RSC or RLC, these costs are placed under Switching Costs. The Remote-to-Host trunk costs for the last four service areas (K–N) could not be separated from the Distribution Link Costs, so they are included under that column.

5. Note that Service Area F is an outlier in the data, in that 2,350 lines were constructed for 346 households. Subsequent research uncovered a prison in this service area, which is in part responsible for the overbuild. We have retested the model without this service area, and the results are essentially unchanged. We have been unable to go back to the sources of the data to investigate this (and some other less obvious) anomolies further.

6. For example, T is the total network cost and L is the number of lines built. Using numbers from Service Area A, T(observed) is $10,545,661 (U.S.) and the variable L is 2,8132 access lines. If the independent variable (L = 28,132) is applied to the equation we see T(power) is estimated to be $11,372,780 (U.S.).

with $R^2 = 0.922$ and a standard error of 0.188. We let this be the network cost function for the incumbent LEC, *Provider 0*. We computed the average cost per subscriber by summing the total network costs in Table 3.1a and 3.1b, and then dividing by the total number of households, yielding an average cost per subscriber of U.S. \$902.36 for our combined study region.

To estimate the network cost function for the new entrant (*Provider 1*), we used Reed's (1993) study. The lowest average cost per subscriber was estimated for the case of an integrated network for telephone and distributed video services with penetration rates of 20% and 40%, respectively. For this case, the average cost per subscriber was U.S. \$1420. We took the ratio of these two average costs (1.57), and use that to differentiate costs of two providers in our interactive game model. We assume that the incumbent *Provider 0* and the new entrant *Provider 1* have cost functions that are the same shape.

SELECTION OF AN ACCESS DEMAND FUNCTION

The demand for access for local telephone service is linked to the demand for usage of service. The "service" can be many varieties and bundles of telecommunication services and features. The question "What price should be charged for access and will the demand for access support that price?" has been extensively debated (Mitchell & Vogelsang, 1991; Schmalensee, 1981). We are more interested in the question, "What is the demand for access?" We briefly list some studies that have examined the issue of demand for access (Table 3.2). The first five studies are reviewed and summarized by Taylor and the later two are newer studies done outside the United States. (Killen, Wynne, & Keogh, 1994; Taylor, 1994; Trotter, 1996). Of the collection, Perl's 1983 study with 1984 revision makes the most extensive use of sociodemographic data, so we chose that model, even though it is 15 years old. We believe that this is still a valid model, as much of the growth of the telephone network has been for second lines, and that the determinants of demand for primary access lines remain basically unchanged

Perl (1983, 1984) used a discrete choice (logit) framework to model a user's choice of whether to have a telephone. This choice is related to monthly cost of service, income, and many sociodemographic factors. His model predicts the logarithm of the odds of having a telephone based on exogenous variables. To focus on access, we selected his model for flat rate access areas, where there is no measured use option available to subscribers. We used Census Tract Household Data from 1990 and matched them to the 14 study service areas (the areas have a population of about 200,000 people in the urban region and 18,000 people in the rural region).

Perl's model estimates *Pr* (the probability of having a telephone) using an intercept (a) and coefficients (b_i) for the vector of sociodemographic characteristics, X_i (Porter 1983) .

$$\ln\left[\frac{P_r}{1-P_r}\right] = a + \sum b_i X_i$$

In our use of the model, the values of all of the socio-demographic characteristics are preset from Census data except the value for the flat rate price parameter (b_f). The equation becomes:

$$\ln\left[\frac{P_r}{1-P_r}\right] = y + \sum b_f X_f$$

where y is a compressed intercept of all the known parameters times their coefficients, plus the Perl intercept. The expression can also be written as:

$$P_r = \frac{1}{1+e^{-(y+b_f X_f)}}$$

which produces the probability of subscribing for the study service area or the penetration rate.

We use this probability function to develop revenue functions for the reaction functions that are used in the interactive game model.

The reaction functions in the game model determine market quantities and market price. To complete the picture, we rewrite Perl's equation to create an inverse demand function:

$$F(q_0 + q_1) = \frac{-1}{b_f\left[\ln\left(\frac{P_r}{P_r-1}\right) - y\right]}$$

Here, q_0 and q_1 are the number of access lines for *Provider* 0 and *Provider* 1, b_f is the coefficient for flat rate price in the Perl model, y is the compressed intercept, and $F(q_0 + q_1)$ is the price for access. The game model that is discussed in the next section selects the quantities that will be produced under described market conditions.

TABLE 3.2
Summary of Access Demand Studies

Study Author	Dependent Variable	Price Elasticity on Basic Access	Type of Access
Allen	main-stations	−.17	cross-section US cities
Feldman	main-stations	−.05	cross-section US cities
Perl	telephone-availability	−.08	cross-section US households
Rash	main-stations	−.11	time-series, Ontario-Quebec
Waverman	main-stations	−.12	time-series, Ontario-Quebec
Wynne	residential-lines	n.a.	time-series, Dublin Ireland
Trotter	residential-lines	−.1	cross-section, Hull UK

DESCRIPTION OF THE APPLIED GAME MODEL

The game model in our application is an implementation of the Green and Porter (1984) model for noncooperative implicit collusion under imperfect information (Bierman & Fernandez, 1993). The Green and Porter model represents the strategic behavior of players through repeated games, which can be used to model the long-term relationships of economic or political situations. The model can represent situations where implicit trust can replace explicit contracts to produce higher expected values for all the players involved. New equilibrium outcomes are possible because the players can condition their decision on information that they received from the previous stages of the game. In the Green and Porter model, the publicly observed information is market price. See Table 3.3.

TABLE 3.3
Intercepts for Different Service Areas

Service Area	y	Service Area	y
A	4.730	H	4.714
B	6.330	I	4.779
C	5.411	J	4.506
D	5.557	K	5.263
E	5.654	L	4.960
F	5.153	M	5.067
G	5.872	N	4.682

The Green and Porter model is well suited to the U.S. local access market because of the present high cost of entry. Also, the product (local access) is homogeneous, and can be offered in one region by two providers. Providers can be collocated, and are assumed to offer facilities-based competition. Information about the industry, such as monthly price for access service, is public information and changes in price are publicly known for residential customers. Providers do not know what facilities their competitors have put into their networks, so they cannot observe one another's outputs (available access lines) directly.

The outcome of the Green and Porter game has also been referred to as a self-enforcing cartel agreement, because the game has many equilibria in which the players can choose to cooperate or to not cooperate. Because the cooperative choices can be self-enforcing equilibria, the participants behave as cartel members even though they are oligopolistic competitors. The participants make inferences about the behavior of the other firm by observing the market price. If the market price remains above a certain value, called the *trigger price*, the firm does not infer a defection from the implicit collusive agreement, but if the market price drops below the trigger price, then retaliation may be warranted.

The players in our model choose a quantity to produce in each period of the game. The Pareto optimal equilibria are limited to a set of strategies in which players choose the Cournot quantities.[7] The total market quantity produced results in a market price, which serves as information to the firms when they choose a quantity to produce in the next time period.[8] In cooperative periods, players can select a smaller quantity, which results in a higher discounted profit.

The interactive game begins in a cooperative period. The firms would be best off if they maintained cooperative behavior so they could extract the highest price. However, outside forces, such as a temporary flux in demand, a demand shock, or one of the firms changing their quantity, can cause changes in the market price. If the market price falls below a certain trigger price, then all firms begin to produce at a higher quantity level.

In an ex post evaluation of how this game actually evolves, all players correctly infer that their rival chose the cooperative quantity in the last period, and that price is low because of demand shock. Punishment then follows automatically as a self-enforcing reaction to the low realized demand. Because, ex ante, players chose an adequate punishment and stuck to it, regardless of whatever the reasons for the price drop, ex post, no cheating takes place. It is optimal to punish only because players do not know what caused the drop in price. Under this uncertainty, players cannot abuse the implicit agreement. During punishment periods, players will produce at higher quantities and accept lower prices, because otherwise all firms would have incentive to defect during cooperative periods, and cooperative quantities would never be enacted.

By applying this quantity game we can find the optimal trigger strategies of the two players. That is, we look for the 4-tuple, of trigger price (tp), number of punishment periods (T), noncooperative equilibria quantity (s_i), and the response quantity $(R_{TP}(q_j))$ which is the optimal output of one player in response to the optimal output of the other player. The trigger strategy $(tp,\ T,\ s_i,\ R_{TP}(q_j))$ for Player i maximizes Player i's expected present and future discounted value. Similarly, for Player j the trigger strategy $(tp,T,s_j,R_{TP}(q_i)t)$ maximizes Player j's expected present and future discounted value.

In our model the expected present discounted value functions for two firms in the market are:

---

7. An equilibrium is Pareto optimal if no other equilibria are Pareto superior to it. An equilibrium is Pareto superior to another if none of the players are worse off in the new equilibrium and one at least is better off than in the other.

8. A quantity game rather than a price game was selected because once the quantity is produced (number of lines), the provider cannot change the price asked for, but must accept the market price. Distribution of profits cannot be renegotiated. This is a condition in a game where no explicit communication is supposed to take place.

$$VC_i(q_i, q_j) = \frac{\pi_i(s_i, s_j)}{1-\delta} + \frac{\pi_i(q_i, q_j) - \pi_i(s_i, s_j)}{(1-\delta) + (\delta - \delta^T)F(x)}$$

$$VC_j(q_i, q_j) = \frac{\pi_j(s_i, s_j)}{1-\delta} + \frac{\pi_j(q_i, q_j) - \pi_j(s_i, s_j)}{(1-\delta) + (\delta - \delta^T)F(x)}$$

where

$$F(x) = \frac{2}{1 + e^{-\ln 3\gamma}} - 1$$

and

$$\gamma = \left(t_p / mp(q_i, q_j) \right)$$

is the probability that the expected market price (*mp*) is greater than the trigger price (*tp*).[9] Other notation is defined as follows.

- s_i and s_j are the strictly noncooperative Cournot quantities that the two firms can choose.
- δ is the discount factor, the way in which firms value future profits.[10]
- T is the number of reversion periods in months.
- tp is the trigger price in U.S. dollars.
- q_i, q_j are the response quantities that optimize the value statements.
- $\pi_i()$ and $\pi_j()$ are profits using the inverse demand function and costs described earlier.

SUMMARY AND DISCUSSION OF MODEL RESULTS

Our model resolves the discounted value statements presented in the previous section. We search for Pareto optimal values over a range of trigger price values (*tp*) for a given reversion length (*T*). When the Pareto optimal discounted value is

---

9. We take this probability function from Bierman and Fernandez (1993). Our results are not particularly sensitive to this function—we tried several and the basic results still hold.

10. We selected a high discount value to help demonstrate potential cooperative outcomes. (i.e., .998). In reality, firms in the local access market may make business decisions with a look ahead of only 5, 7, 13, or 18 years. This would have a discount factor of .35, .50, .70, or .75, respectively. An alternative selection of δ would change the model's outcomes.

found, at a particular tp, we then run the model again at that tp over a range of different reversion lengths T. We then repeat the process, testing discounted values at different tp using the new T. The process is repeated until no higher discounted values can be found. From this procedure we compute the best response quantities, $R_{TP}(q_j)$ and $R_{TP}(q_i)$. We also verify the same response quantities using two different root-finding functions in Mathematica.

Figure 3.1 is an example of the Pareto optimal response quantities computed for Service Area A. In the figure, the first crosspoint nearest to the axis origin represents the best response quantities for each firm and produces the highest discounted values; it is also the Nash equilibrium. The next crosspoint shows the Cournot quantities, which produce lower discounted values. The furthest crosspoint from the origin shows another suboptimal equilibrium point where selected quantities produce still lower discounted values.

FIGURE 3.1
Response curves for Service Area A.

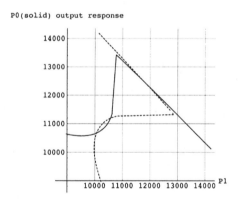

Result 1: The Pareto Optimal Reversion Length Is Long

After examining all 14 service areas, we found that in most of the service areas the discounted values of both firms always increased, but with decreasing increments, as we increased the reversion length (T) in months. The model also showed many suboptimal tuples where $T < \infty$. In these cases, the collusive quantities still produced higher expected values than Cournot quantities, so selecting these collusive quantities was superior to selecting the Cournot quantities. In game theory, the outcomes produced by the suboptimal strategies would not be part of the core, because players can do just a little better by punishing forever. However, in practice, firms may not be concerned with the very small difference in expected discounted values that result from using a finite reversion length. In summary, the response curves, of which Figure 3.1 is an example, are plots of the numerical values of the best quantity responses, and demonstrate the point. The quantities at the intersection closest to the origin is the Nash equilibrium in these tringger price strategies, where the discounted expected value for both firms is maximized.

Result 2: Trigger Prices Are Linked to the Sociodemographic Factors

We can also comment about the resulting trigger prices for the 14 service areas. First, the cooperative response prices that can be derived from the computed cooperative response quantities are always greater than the trigger prices. The trigger prices themselves are related to the y intercept that was computed for each service area using the Perl model (see Tables 3.3 and 3.4). The trigger price is higher in areas that would support greater demand for access. In Table 3.5 we list the expected market prices that were computed using the quantities in Table 3.4.

Result 3: Universal Service Supports Are Desirable

An important result of this study is that universal service supports that would promote greater service penetration in these study areas is desirable because collusive behavior, even if implicit, can occur. Such behavior will adversely affect the number of households receiving service due to higher access prices (see Tables 3.1a, 3.1b and 3.4). We note the total number of households served in each service area using Cournot quantities or cooperative quantities. In Service Area A this is 21,765 households using Cournot quantities and 20,741 using cooperative quantities and amounts to service penetrations of 82.38% and 78.51%, respectively.[11] The number of households served in a market such as this would be well below current expectations for universal service.

This result supports the notion that some form of subsidy is necessary if social goals are to be achieved, even though this is not an explicit result of the model.[12] Our data were for capital costs only, and do not include operating costs, suggesting that the actual cost basis is higher than the data reported here suggest. These higher costs could be expected to result in still higher prices, which would in turn reduce service penetration further.

---

11. We assume that all costs for the network are recovered out of the access cost. In practice, a portion of the costs may be recovered from usage-based costs, which would improve the service penetration.

12. This issue is actually a good bit more complex. For example, we do not compute potential revenues from other services produced over the same infrastructure.

Implications for Local Exchange Competition

Our model describes conditions in which collusive behavior in a local exchange market can occur, even though the firms are in fact competitors. We focus our discussion on several important conclusions.

- **Collusive behavior could occur in the study area, but tested conditions are very strict.**
  Under smaller discount factors (δ) the reversion length (T) must grow larger to find collusive outcomes and in some cases there is no collusive outcome (we did not investigate the minimal conditions, the smallest reversion lengths (T), to produce collusive response).
- **Competition is sustainable even in areas where a grim strategy is Pareto optimal.** The grim strategy does not mean that one firm would lose money and therefore exit the market, but instead implies that in order to sustain the collusive price, providers must punish forever. In this case, the only other price the providers can obtain is the Cournot price, which must exceed the long run average. Our market cost data in Table 1a and Table 1b show that an average cost function is downward sloping and approaching a constant marginal cost. Therefore, a price above marginal cost would ensure a price above long-run average costs. Because we know the Cournot price is above marginal cost we can conclude the firms would not go out of business by not colluding, they would just earn lower economic profits.

TABLE 3.4
Trigger Strategies, T = 5,000

SA	T	tp	Cournot Q_0	Response Q_0	Cournot Q_1	Response Q_1	Discounted Value P_0	Discounted Value P_1
A	1227	53	11196	10679	10569	10062	324329671	273652415
B	1697	73	6981	6831	6639	6491	281849026	243035603
C	5000	61	6996	6760	6577	6345	231823833	192891963
D	5000	63	4398	4263	4075	3943	145638897	115581710
E	5000	68	636	617	492	474	16447051	6079643
F	5000	66	132	125	132	125	888902	888902
G	1245	69	1596	1552	1408	1365	51235412	34118957
H	5000	57	781	743	573	539	16017613	4297454
I	5000	58	771	733	568	534	16088010	4461609
J	5000	55	738	700	516	482	13990323	2548811
K	5000	62	1019	982	832	797	26044054	12617863
L	5000	60	660	632	477	452	14047019	3361014
M	5000	61	745	714	564	535	16846068	5481925
N	1860	59	531	503	329	305	9794068	112343

- **Low service penetration is possible without universal service supports.** This study indicates that the percentage of households that would receive local exchange access under market conditions is well below the expected penetration rates for universal access. Along with lower penetration rates, the price for access would be much higher than typical local access tariffs in the United States.

  This result supports proposed policies for universal service subsidies. More specifically, this model uses a demand for access model that represents a study area's sociodemographic features. The sociodemographic y intercept for all the 14 study areas reflects a strong demand for telephone access. The model shows areas with low demand for access would have lower trigger prices and that collusion could more easily occur in those areas. From this observation we can conclude that universal service support would be particularly important in areas with lower demand for access.

TABLE 3.5
Expected Market Price, in Monthly U.S. Dollars

SA	Cournot Price	Cooperative Response Price
A	67.39	72.61
B	92.89	96.34
C	78.22	82.66
D	81.04	85.16
E	87.16	90.85
F	84.21	88.70
G	87.67	91.47
H	73.12	77.89
I	74.07	78.96
J	70.79	75.59
K	79.71	83.92
L	87.62	90.75
M	59.25	67.97
N	75.10	80.04

- **Our model cannot be applied when there are dynamic changes in the demand for access.** The application of our trigger price model is appropriate for the scope of this analysis because the demand for access has been growing at a steady and predictable rate for the past decade. We start with an initial state of demand and make projections forward based on that initial state. New technology, such as wireless access, can introduce lower cost access networks and substitute products and services.

We conjecture that non-facilities-based entry into the local access market creates less clear market boundaries between service markets. Because access services can be more easily combined with different bundles of services, it becomes difficult to estimate demand for access alone. We also speculate that the policies for allocating loop costs based on presubscribed lines would be less effective. Inter-exchange carriers (IXCs) could pass on better savings to partner LECs where partnerships may only evolve in higher demand areas. Again, market pressure that will lower consumer prices we predict would not arise easily. This points again to the present applicability of our model and the need for universal access assistance for consumers.

REFERENCES

Bellcore. (1994). *BOC Notes on LEC Networks*. (No. SR-TSV-002275, Issue 2).

Bierman, H. S., & Fernandez, L. (1993), *Game theory with economic applications*. Boston: Addison-Wesley .

Green, E., & Porter, R. (1984). Noncooperative collusion under imperfect price information. *Econometrica, 52*, 87–100.

Hatfield Associates, Inc.(1994) *The cost of basic universal service*. MCI Communications Corporation. Boulder, CO.

Killen, L., Wynne, L., & Keogh, G. (1994) *Estimation of demand for residential telephone connections in Ireland*. (Tech. Rep. No. CA-0594). Dublin, Ireland: Dublin City University.

Mitchell, B.M. (1990) *Incremental capital costs of telephone access and local use*. (Tech. Rep. No. R-3909-ICTF). Santa Monica, CA: The RAND Corporation.

Mitchell, B. M., & Vogelsang, I. (1991). *Telecommunications pricing: Theory and practice*. New York: Cambridge University Press.

Molka-Danielsen, J., & Weiss, M. B. H. (1996). Local telephony access costs: Empirical cost functions for rural and urban areas. In *Proceedings of the 4th International Conference on Telecommunications Systems, Modeling and Analysis* (pp. 21–24).

Perl, L.J. (1983). *Residential demand for telephone service*. (Tech. Rep.). White Plains, NY: National Economic Research Association, Inc.

Perl, L.J. (1984). *Revision to NERA's residential demand for telephone service 1983*. (Tech. Rep.). White Plains, NY: National Economic Research Association, Inc.

Porter, R. (1983). Optimal cartel trigger price strategies. *Journal of Economic Theory, 29*, 313–38.

Reed, D. (1993). The prospects for competition in the subscriber loop: The fiber-to-the-neighborhood approach. Paper presented at the Twenty-First Annual Telecommunications Research Policy Conference, Solomons Island, Maryland.

Schmalensee, R. (1981). Monopolistic two-part tariff arrangements. *Bell Journal of Economics, 12* (pp. 445-466).

Taylor, L.D. *Telecommunications demand in theory and practice.* Dordrect, Netherlands: Kluwer Academic.

Trotter, S. (1996) The demand for telephone services. *Applied Economics, 28,*175–84.

The Implications of By-Pass for Traditional International Interconnection

Douglas Galbi
International Bureau, FCC

An important decision that international telecommunications carriers make is how much traffic to send under traditional international interconnection arrangements (settled traffic) and how much to send under new alternative arrangements (by-pass traffic). This chapter presents an equilibrium model of home and foreign carriers' optimal routing choices for international traffic. The model suggests that movements in by-pass prices are likely to play a dominant role in determining the welfare implications of traditional international interconnection arrangements.

Despite the vast amount of work on the economics of interconnection for telecommunications networks,[1] there has been relatively little analysis of international interconnection arrangements.[2] This is understandable, for it is not obvious why international interconnection arrangements should be distinguished from domestic arrangements. From a global welfare perspective, efficient international inter-

---

1. See Armstrong, Doyle, and Vickers (1995), Baumol and Sidak (1994), Kahn and Taylor (1994), Katz, Rosston, and Anspacher (1995), Laffont and Tirole (1994), Mueller (1997), Ralph (1996), and Tye (1994), among others.

2. A small body of academic work has analyzed these arrangements and alternative regulatory rules for international interconnection. Kwerel (1984, 1987) considered the effect of increasing competition among U.S. international carriers and Stanley (1991) described the problem of ballooning net U.S. international interconnection payments. O'Brien (1989) provided an interesting theoretical analysis of the impact of requiring uniform interconnection rates among U.S. carriers. Other economic models, which also focus on pricing issues, include Hakim and Lu (1993), Carter and Wright (1994), Cave and Donnelly (1996), and Yun, Choi, and Ahn (1997). Alleman, Rappoport, and Stanley (1990) analyzed accounting rate reform possibilities, and Alleman and Sorce (1997) and Galbi (1997) provide further work in this area

connection would entail an open, competitive market for international communications bandwidth and an efficient domestic interconnection regime that did not distinguish between domestic and international minutes. There would be no separate regulatory regime for international interconnection.

The history of international communications has, however, traced a distinctive institutional path.[3] National carriers conceptualized international telecommunications as a jointly provided service. They agreed on collection rates (retail prices), the revenue from which was to be shared equally between the two carriers providing the service. When collection rates (retail prices) diverged in response to country-specific factors, carriers continued to share equally rates that were then called *accounting rates*.[4] International interconnection thus evolved as a system of bilaterally agreed mutual (i.e., equal), termination rates. These mutual termination rates are traditionally called *settlement rates*, reflecting linguistically their "accounting" heritage.

The development of multiple international carriers in the United States led to new rules to support settlement rates. The Federal Communications Commission (FCC) established rules that required uniform settlement rates among U.S. international carriers and divided the market for terminating foreign-billed international traffic among U.S. carriers, on a country-by-country basis, in proportion to a U.S. carrier's share of domestic-billed international traffic with the corresponding country.[5] Other countries that have licensed multiple international carriers have sanctioned similar rules or are considering adopting such rules. Such rules are called *proportional return rules*, and international traffic that flows under such rules is known as *settled traffic*.

New opportunities for by-passing traditional settlement arrangements are rapidly emerging. Under the World Trade Organization (WTO) agreement, 52 countries have committed to liberalize opportunities for carriers to provide public switched voice international service independent of traditional settlement arrangements.[6] More generally, the broad-based move to competition in telecommunications in countries around the world, the convergence of different media to digital signals, the

---

3. See Ergas and Patterson (1991) for a discussion of the historical development of international interconnection.

4. There is an economic logic for the persistence of this institution (see Galbi, 1997).

5. See Implementation and Scope of the International Settlements Policy for Parallel Routes, CC Docket No. 85-204, 51 Fed. Reg. 4736 (Feb. 7, 1986), recon. 2 FCC 1118 (1987), further recon. 3 FCC Rcd 1614 (1988); Regulation of International Accounting Rates, 6 FCC Rcd 3552 (1991), recon. 7 FCC Rcd 8049, Fourth Report & Order, CC Docket No. 90-337, Phase II. This latter order outlines circumstances and means by which the FCC will waive its international settlements policy.

6. More specifically, these countries have committed to allow public international switched voice service over resold private lines interconnected at both ends to the public switched network.

growth of the global Internet, and the rapidly falling price of digital signal processing are increasing the number of technical and market possibilities for providing international voice communications. Internet telephony epitomizes these trends.

Alternative routing opportunities make traffic routing a strategic choice for carriers. This chapter considers the implications of by-pass for traditional settlement
arrangements. The first section analyzes, for a given volume of settled traffic from
foreign carriers, the routing decisions of home carriers. The following section considers equilibria that encompass both home and foreign carrier routing choices.
These equilibria show that flows of settled traffic may persist even when by-pass
prices in both directions are below the settlement rate. As the final two sections
emphasize, the welfare implications of a given settlement rate depend significantly
on by-pass prices. By-pass prices change rapidly in response to markets and technologies whereas settlement rates are typically negotiated within a sphere of government participation or oversight, with the deliberate speed characteristic of
bureaucracies. Policymakers should thus recognize the limitations of their ability
to manage through proportional return and settlement rate policies the division of
welfare in international telecommunications.

HOME CARRIERS' ROUTING REACTION FUNCTIONS

This section considers equilibria among n competing home carriers that send international traffic to a foreign country either as settled traffic under proportional
return or as by-pass traffic (for which proportional return does not apply). Although international carriers make a variety of important economic choices, the
analysis focuses on cost-minimizing traffic allocations, given traffic volumes and
international interconnection prices. Such a focus highlights the implications of
by-pass for traditional settlement arrangements.

The model is defined as follows. The prices that home carriers face for sending
home-billed settled traffic and by-pass traffic are p_a and p_b^F, respectively, whereas
the costs that home carriers incur for handling foreign-billed settled and by-pass
traffic are c_a^H and c_b^H respectively. For Home Carrier i, T_i^H is total home-billed
international traffic and s_i^H is the share of home-billed traffic sent as settled traffic.
Then Home Carrier i's net international interconnection expense is

$$X_i^H = p_a s_i^H T_i^H - p_a \left(1 - \frac{c_a^H}{p_a}\right) \frac{s_i^H T_i^H}{\Phi^H} \Phi^F + p_b^F \left(1 - s_i^H\right) T_i^H \tag{1}$$

where Φ^H and Φ^F are total settled traffic from the home and foreign countries, respectively.

The terms in Equation 1 have straightforward meanings. The first term represents the total interconnection expense for outgoing settled traffic, the second term

represents profit from incoming settled traffic received under proportional return, and the third term represents the expense for outgoing by-pass traffic. Profit from terminating foreign carriers' by-pass traffic is not included in Equation 1. This can be interpreted in two ways. If the amount of foreign by-pass traffic that Carrier i terminates does not depend on the share of settled traffic it sends, then the former quantity does not affect the optimal choice of the latter. Alternatively, one can assume that the market for by-pass is competitive ($p_b{}^H = c_b{}^H$), so that carriers earn only a normal return on by-pass termination.[7]

Carriers route international traffic so as to minimize their net international interconnection expenses. Carrier i's marginal cost of sending a minute of traffic as settled traffic rather than as by-pass traffic is

$$\frac{dX_i^H / ds_i^H}{T_i^H} = p_a\left[1 - \frac{p_b^F}{p_a} - \frac{\Phi^F \Phi_{\neq i}^H}{(\Phi^H)^2}(1 - \frac{c_a^H}{p_a})\right] \qquad (2)$$

where $\Phi_{\neq i}^H$ is settled traffic sent by all home carriers except Home Carrier i. Note that Carrier i, such that i maximizes $\Phi_{\neq i}^H$, is the carrier with the smallest volume of settled traffic. Equation 2 thus shows that the carrier with the smallest amount of settled traffic has the lowest marginal cost of shifting traffic to the settlements regime. Note that the marginal cost to a particular carrier of sending settled traffic depends on how much settled traffic the other carriers send. In order to minimize Equation 1, each carrier chooses s_i^H such that Equation 2 equals zero.

Consider Nash equilibria with interior solutions to Equation 2 for $i = 1, \ldots, n$.[8] Rearranging Equation 2 shows that

$$\frac{\Phi_{\neq i}^H}{\Phi^H} = \frac{\Phi^H}{\Phi^H r^H} \quad where \quad r^H = \frac{(1 - \frac{c_a^H}{p_a})}{(1 - \frac{p_b^F}{p_a})} . \qquad (3)$$

Because the right side of Equation 3 does not depend on i, $\Phi_{\neq i}^H / \Phi^H = (n-1)/n$. Hence total settled traffic sent from the home country, Φ^H, is

$$\Phi^H = \frac{n-1}{n} r^H \Phi^F . \qquad (4)$$

---

7. Note that handling by-pass traffic may be more costly than handling settled traffic.

8. This means that $s_i^H < 1$ for all i. As will be apparent subsequently, if $\Phi^F > 0$, there are no solutions with $s_i^H = 0$ for some i.

This Nash equilibrium among home carriers implies a *threshold routing rule*. Home carriers, which may have different total traffic volumes to a given country, choose their settled traffic shares so that each carrier sends an equal amount of settled traffic. In particular, in an interior solution, each home carrier sends Φ^H/n minutes of settled traffic, where Φ^H is given by Equation 4. Additional traffic $T_i^H - \Phi^H/n$ is sent via by-pass. The intuition for this result is the following. Because of proportional return, the cost of switching a minute of traffic from a by-pass route to the settlement system depends on a carrier's share of total settled traffic. Because the cost of by-pass is the same for all carriers, there must be symmetry in the shares of settled traffic.

Carrier reaction functions may also involve boundary solutions. As the volume of foreign settled traffic increases, home carriers will push to the boundary $s_i = 1$ in order from smallest to largest T_i^H. Given that some carriers choose to send all traffic as settled traffic, Equation 3 still holds for carriers with $s_i^H < 1$, and carriers that send some by-pass traffic all send an equal amount of settled traffic. The effect of boundary solutions is to make Φ^H concave as a function of Φ^F.

Consider the case where there are two carriers, one of which sends all its traffic as settled traffic.[9] Let $\Phi_1^H = T_1^H < \Phi_2^H < T_2^H$ for Home Carriers 1 and 2. Then Equation 3 implies

$$\frac{T_1^H}{\Phi^H} = \frac{\Phi^H}{\Phi^F r^H} \ for \ \Phi^H > 2T_1^H. \tag{5}$$

Solving Equation 5 for home settled traffic gives

$$\Phi^H = \sqrt{r^H T_1^H \Phi^F} \ for \ \Phi^H > 2T_1^H. \tag{6}$$

The amount of settled traffic that Home Carrier 2 sends is $\Phi_2^H = \Phi^H - T_1^H$.

There are several important general properties of the routing functions in Equation 4 and Equation 6. First, new carriers have a large incentive to send some settled traffic, and, all else held constant, the volume of settled traffic increases as the number of carriers increases. More significantly, carriers send some settled traffic even if a by-pass option is available at a lower price than the price for settled traffic. The amount of settled traffic sent is a continuous function of the by-pass price, the settlement rate, the cost of handling settled traffic, and the amount of settled traffic received.

---

9. The more general case is easily solvable, but will not lead to a closed form expression for the Nash equilibrium in the subsequent section.

NASH EQUILIBRIA WITH HOME AND
FOREIGN CARRIER ROUTING CHOICES

Home and foreign carriers' optimal routing choices are in general interdependent. Proportional return simplifies the interdependency: home carriers' routing costs depend only on the total amount of settled traffic from the foreign country, and the analogous property holds for foreign carriers' routing costs. Thus the analysis of foreign carriers' routing choices, for a given volume of home country settled traffic, is analogous to the analysis in the previous section. The overall Nash equilibrium requires a solution for Φ^H and Φ^F consistent with both home and foreign carriers' reaction functions.

Consider first the case where there is a single foreign international carrier. If its by-pass price is lower than the settlement rate, its dominant strategy is to send all its international traffic as by-pass traffic. Given this strategy, if home carriers also face a by-pass price less than the settlement rate, they will also send all their traffic as by-pass traffic. Thus a bang-bang solution, with respect to movements in by-pass prices, can emerge for a Nash equilibrium with a foreign monopolist.

Suppose, however, that a foreign monopolist can commit to a particular routing strategy and hence can act as a Stackelberg leader in the routing game. Such a situation may reflect, for example, the bureaucratic inertia in the decision making of a state-owned telecommunications carrier in contrast to the decision-making process in competitive private carriers. The foreign monopolist's net international interconnection cost, given the home-country reaction function in Equation 4, is

$$X^F(s^F) = p_a s^F T^F - p_a (1 - \tfrac{c_a^F}{p_a}) \tfrac{n-1}{n} r^H s^F T^F + p_b^H (1 - s^F) T^F. \qquad (7)$$

Thus the foreign monopolist's marginal cost of sending a minute as settled traffic rather than as by-pass traffic is

$$\frac{dX^F / ds^F}{T^F} = p_a \left[1 - \tfrac{p_b^H}{p_a} - \tfrac{n-1}{n} r^H (1 - \tfrac{c_a^F}{p_a}) \right]. \qquad (8)$$

The right side of Equation 8 is some constant K. If K > 0, the foreign carrier will send all traffic as by-pass traffic, and so will the home carriers. If K < 0, the foreign carrier will send some traffic as settled traffic.[10] If the monopolist sends settled traffic, it follows a threshold routing rule, whereby all traffic below a threshold value M is sent as settled traffic and any traffic above that threshold is sent as by-pass

---

10. Note that if $p_b^F > c_a^F$ and $p_b^H > c_a^H$ then for sufficiently large n, K < 0.

traffic.[11] Moreover, the threshold changes continuously in response to changes in by-pass prices, settlement rates, and costs, and hence routing choices do not change abruptly in response to changes in by-pass prices.

With multiple home and foreign carriers there are three possible types of Nash equilibria. One is a simple no-alternative scenario: With by-pass prices higher than settlement rates, all traffic is passed as settled traffic. With by-pass prices lower than settlement rates, there is another simple equilibrium: All traffic is passed as by-pass traffic. However, these two routing equilibria do not exhaust the possibilities; a third equilibria, featuring some settled traffic and some by-pass traffic, is also possible.

Consider the case where there are n home carriers and m foreign carriers, each of which handles settled traffic according to proportional return.[12] The home country settled traffic reaction function is given by Equation 4. The analogous reaction function for the foreign country can be written as

$$\Phi^H = \frac{m}{(m-1)r^F}\Phi^F.$$

(9)

Because the home country reaction is convex in Φ^F and the inverse of the foreign country reaction function is concave in Φ^F, an interior equilibrium exists if and only if

$$\frac{m\,n}{(m-1)(n-1)} < r^F r^H.$$

(10)

Assume that settled traffic is handled efficiently. Thus the price for by-pass, which may involve additional technology and/or additional business and regulatory risk, can be no lower than the cost of handling settled traffic. Hence $r^H > 1$ and $r^F > 1$. Thus Equation 10 holds when there are sufficiently many home and foreign carriers.

Now assume that there are only two foreign carriers ($m = 2$), and in equilibrium Foreign Carrier 1 sends all its traffic as settled traffic. This adds a concave segment

---

11. If there are n equal-sized home carries, then $M = \frac{n\Sigma^H}{(n-1)r^H}$, where Σ^H is total international traffic from the home market.

12. Carrier-by-carrier proportional return is a decentralized approach for ensuring aggregate proportional return. Other approaches to ensuring aggregate proportional return require move intensive coordination of traffic flows among carriers. Without a publicly known mechanism for producing aggregate proportional return, a carrier does not know which parties to hold responsible if it does not receive an appropriate traffic allocation.

to the foreign country reaction function that is analogous to that given in Equation 6. It can be written as

$$\Phi^H = \frac{(\Phi^F)^2}{r^F T_1^F} \ for \ \Phi^F > 2T_1^F. \tag{11}$$

Figure 4.1, which plots the home and foreign country reaction functions, shows a possible Nash equilibrium given sufficient total traffic volumes from each home carrier, that is, interior solutions for all home carriers.[13] Solving Equation 11 and Equation 4 for the Nash equilibrum gives

$$\Phi^H = \frac{n-1}{n} r^F r^H T_1^F, \Phi^F = \left(\frac{n-1}{n} r^F\right)^2 r^H T_1^F. \tag{12}$$

FIGURE 4.1
Home and Foreign Countries' Reaction Functions (RF)

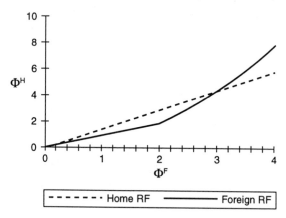

The traffic ratio is

$$\frac{\Phi^H}{\Phi^F} = \frac{n}{(n-1)r^F}. \tag{13}$$

---

13. Equation 10 is assumed to hold.

In terms of the effects of successive optimal reactions, this equilibrium is stable, whereas the "no settled traffic" equilibrium is not stable. Suppose, for example, that foreign settled traffic is slightly above the rightmost intersection of the reaction functions in Figure 4.1. Home carriers' optimal reaction to this level of foreign settled traffic is a level of home settled traffic that in turn implies, via the foreign carriers' reaction function, a lower level of foreign settled traffic. In contrast, consider the equilibrium where no carrier sends any settled traffic. If, from that position, some carrier sends some settled traffic, all other carriers have an incentive to send some settled traffic as well. Thus settled traffic flows move further away from the "no settled traffic" equilibrium. Particularly given the historical existence of significant settled traffic flows, these dynamics suggest that the "no settled traffic" equilibrium is much less likely than the equilibrium with some settled traffic.

An important feature of this latter equilibrium is that ratio of home and foreign settled traffic depends on the settlement rate, the cost of handling settled traffic, the foreign by-pass price, and the number of home carriers. The total volumes of traffic sent from the home and foreign countries, which might be considered structural aspects of the market, have no effect on the equilibrium balance of settled traffic. Moreover, increased competitiveness of the home market (larger n) leads to a lower settled traffic ratio. These results at least suggest that policy analysis neglecting the role of by-pass in shaping the balance of settled traffic overlooks crucial aspects of carriers' routing strategies.

WELFARE ANALYSIS

The threshold routing rules derived previously simplify the welfare analysis of international interconnection. Because those rules imply that by-pass prices represent the marginal cost of interconnection (for carriers sending some by-pass traffic), profit-maximizing international carriers will set retail prices based on prevailing (market-driven) by-pass prices. Settlement rates and net settlement payments under the traditional international interconnection arrangements affect only carriers' average costs. Thus in the short run (no entry or exit), settlement rates and net settlement payments affect only the international division of producer surplus, not retail prices or consumer welfare.

Analyzing the international division of producer surplus requires some standard for comparison. In this section the standard of comparison will be taken to be net international interconnection expenses with international traffic terminated at the cost, which may vary among countries, of terminating settled traffic. The cost of terminating settled traffic is taken to represent the least cost means for handling settled traffic. If international traffic was terminated at this price, there would no role for by-pass. Moreover, retail prices would reflect this interconnection price rather than a by-pass price, and there would be consumer welfare effects in the short run (without entry or exit). These consumer welfare effects are largely ignored in the following analysis.[14]

The home country would prefer to shift to cost-based international interconnection if doing so lowered net international interconnection expenses. Thus the home country would want to shift to cost-based rates if

$$c_a^F \Sigma^H < p_a \Phi^H + p_b^F B^H - (p_a - c_a^H)\Phi^F - (p_b^H - c_b^H)B^F \qquad (14)$$

where Φ^H and B^H are, respectively, settled traffic and by-pass traffic sent from the home country, and $\Sigma^H = \Phi^H + B^H$ is total home country international traffic. An analogous equation applies for the foreign country. Rearranging terms gives

$$c_a^F \Sigma^H - c_a^H \Sigma^F - (c_b^H - c_a^H)B^F < p_a(\Phi^H - \Phi^F) + p_b^F B^H - p_b^H B^F. \qquad (15)$$

The right side of Equation 15 represents the home country's net international interconnection payment under the given system. The first two terms of the left side are net interconnection expenses under cost-based interconnection. With $c_b^H = c_a^H$, the third term on the left side of Equation 15 vanishes and the equation becomes negatively symmetric with respect to the home and foreign countries.

This means that, holding consumer welfare constant, home and foreign country incentives to shift to cost-based interconnection conflict: If one country prefers cost-based interconnection, the other prefers the given system.

Note that a positive net international settlement payment (i.e., a positive first term on the right side of Equation 15) is not sufficient to imply that the home country is better off with cost-based interconnection. The net payment for by-pass traffic, the cost difference for handling settled traffic, and the relationship between settled traffic costs and by-pass costs also matter for the welfare comparison. As others have pointed out, if the costs of handling settled traffic are sufficiently asymmetrical, a country with a positive net settlement payment would experience an even larger net international interconnection payment under cost-based international interconnection.[15]

Moreover, a country might have lower net international interconnection expenses if it abandoned proportional return. Assume that by-pass prices are lower than settlement prices for home and foreign carriers. If the home country abandoned proportional return, its carriers would send all their traffic as by-pass traffic, and

---

14. Thus the analysis essentially assumes that retail prices do not change if average or marginal interconnection costs change. This situation might be considered the very short run.

15. See Walker (1995) and Chowdary (1997).

hence so would foreign carriers. The home country's net international interconnection expenses would fall if

$$p_b^F \Phi^H - (p_b^H - c_b^H)\Phi^F < p_a \Phi^H - (p_a - c_a^H)\Phi^F. \tag{16}$$

Rearranging yields

$$\frac{(p_a - c_a^H) - (p_b^H - c_b^H)}{p_a - p_b^F} < \frac{\Phi^H}{\Phi^F}. \tag{17}$$

Taking $c_a^H = c_b^H$ and letting p_b^H approach p_a from below shows that this equation holds for some parameter values. More generally, Equation 17 shows that a home country price advantage for by-pass favors a shift away from proportional return.

Two factors create opportunities for mutual gains from a movement to cost-based international interconnection. The first is that reducing the marginal costs for terminating international calls will, given profit maximization, lead to a reduction in retail prices, increased traffic flows, and subsequent consumer benefits. Analyzing such benefits requires a more complex model.

A second factor that creates opportunities for mutual gains is that by-pass may be inefficient. Some forms of by-pass traffic involve third-country routing. The extra transit, switching, and managerial coordination required is likely to make handling such traffic more expensive than handling settled traffic under established bilateral arrangements. Other forms of by-pass involve placing voice traffic on alternative networks. The existing international network has been optimized for voice traffic, hence it is likely to be more cost-efficient for such traffic than other alternative networks designed for more diverse traffic streams. In addition, the provision of by-pass services may expose a carrier to significant regulatory and commercial risks. This additional risk raises the cost of by-pass services relative to traditional settled traffic. Reforming the settlement process (making rates more closely aligned with costs, and explicitly sanctioning alternative routing practices) can reduce the inefficiencies associated with by-pass.

NUMERICAL CALIBRATION

This section numerically calibrates the preceding model. For simplicity the analysis focuses on the case of multiple carriers in the home country and a foreign monopolist. Given that countries are only beginning to move toward competitive international service, while many remain with a monopoly international carrier, this case is also empirically relevant.

The calculations assume a settlement rate of US$0.45 ($p_a = 0.45$) per minute. The cost of handling settled traffic is assumed to be US$0.06 per minute in the home country and US$0.09 per minute in the foreign country ($c_a^H = 0.06$, $c_a^F = 0.09$). The higher cost for the foreign country reflects the cost inefficiency of a monopoly provider. The cost of providing by-pass service to the foreign country, c_b^F, is assumed to be US$0.12 per minute. The cost of handling by-pass traffic is assumed to be higher than the cost of handling settled traffic because of the additional arrangements and risks associated with by-pass traffic.

Some additional assumptions are needed about traffic volumes and market structure. Assume that there are twice as many minutes of international traffic from the home country to the foreign country as there are in the opposite direction. Assume that there are four equal-sized carriers providing international service in the home country ($n = 4$). The assumption that the home carriers are equal-sized implies that either all the home-billed traffic is settled traffic or there is an interior solution for the home carriers and Equation 4 holds.

As shown earlier, carriers have an incentive to send some settled traffic even if the by-pass price is lower than the settlement rate. In Table 4.1, the first column gives values for the home carriers' by-pass price. The second column gives the lower bound for the by-pass price such that, for any by-pass price above that bound, the foreign carrier will send all its international traffic as settled traffic. By-pass prices considerably below the settlement rate are consistent with the foreign carrier sending all its traffic as settled traffic. Columns 3 and 4 show the home carriers total by-pass traffic to the foreign country and the ratio of home to foreign settled traffic for the home by-pass price in the first column and a foreign by-pass price above the bound in the second column. With a settlement rate of US$0.45, home carriers exploit by-pass only if the by-pass price falls below US$0.32. As the by-pass price falls further, the home carriers send more by-pass traffic. With a by-pass price just below US$0.16, the home carriers send the same amount of by-pass and settled traffic, and the ratio of home-billed settled traffic to foreign-billed settled traffic is 1. A by-pass price of US$0.16 offers a 33% margin over the cost of handling by-pass traffic. These calculations show that, given proportional return, by-pass prices have to be significantly below the settlement rate to affect the settled traffic ratio. Nonetheless, such by-pass prices appear to be economically feasible.

By-pass prices can significantly affect the home and foreign carriers' net international interconnection expenses. Columns 5 and 6 of Table 4.1 show home and foreign carriers' net international interconnection expenses relative to their expenses with cost-based settlement of international traffic.[16] By-pass reduces the surplus of settled traffic from the home carriers to the foreign carrier, and increases the foreign carrier's relative cost for international interconnection. Note, however, that with a home by-pass price of US$0.16, both the home and foreign carriers

---

16. That is, the exchange of international traffic at the cost of handling settled traffic.

would prefer to shift to cost-based interconnection. An advantage of shifting to cost-based international interconnection is that it avoids the efficiency loss associated with by-pass. With a by-pass price of US$0.16, this loss amounts to 12% of the total cost of efficient, cost-based interconnection.

TABLE 4.1
The Effects of By-Pass Prices

Home By-Pass Price	Foreign By-Pass Price	Home By-Pass Traffic	Settled Traffic Ratio	Home Rel. Net Expense	Foreign Rel. Net Expense	Eff. Cost of By-Pass
0.32	>0	0	2.00	2.83	-4.50	0%
0.30	>0	5	1.95	2.79	-4.35	1%
0.28	>0	28	1.72	2.57	-3.57	3%
0.26	>0	46	1.54	2.35	-2.81	6%
0.24	>0	61	1.39	2.13	-2.07	8%
0.22	>0	73	1.27	1.90	-1.34	9%
0.20	>0.03	83	1.17	1.68	-0.63	10%
0.18	>0.06	92	1.08	1.46	0.08	11%
0.16	>0.09	99	1.01	1.24	0.79	12%
0.14	>0.11	106	0.94	1.01	1.49	13%
0.12	>0.13	111	0.89	0.79	2.18	14%

CONCLUSIONS

By-pass opportunities have important implications for traditional international interconnection. In the simple case of two international carriers negotiating interconnection, by-pass can be interpreted as a nonnegotiated option that bounds interconnection prices. Proportional return, however, creates more complex effects. Under proportional return, carriers may follow threshold routing rules and may rationally send some settled traffic even if by-pass prices are below the settlement rate. Moreover, with multiple home and foreign carriers following proportional return, the balance of settled traffic in a Nash equilibrium depends on by-pass prices but is independent of the overall international traffic flows between the two countries.

More importantly, by-pass makes the welfare implications of traditional international interconnection agreements subject to fast changing technological and market developments. The experience of Bell Atlantic within the United States illustrates how market developments can shift the welfare implications of an interconnection agreement. In interconnection negotiations, Bell Atlantic pushed for relatively high reciprocal interconnection rates, presumably because of past experience that new entrants had a surplus of traffic flowing out from their networks.

With such rates in place, the new entrants then pursued new customers, such as Internet service providers, that had a high ratio of inbound to outbound calls. The result was that Bell Atlantic soon found itself accumulating large liabilities for interconnection payments.[17] With respect to international interconnection, settled traffic ratios are rapidly becoming a function of strategic routing and marketing choices. These developments imply that proportional return and extensive regulation of settlement rates are becoming less useful policy tools.

ACKNOWLEDGMENTS

The views expressed in this chapter are those of the author and do not necessarily reflect the views of the Federal Communications Commission or its staff. I am grateful for comments and suggestions from Jerry Duvall, Joe Farrell, and Tom Spavins.

REFERENCES

Alleman, J. H., Rappoport, P. N., and Stanley, K. B. (1990), Alternative settlement procedures in international telecommunications service. In D. Elixman & K. H. Neuman (Eds.), *Communications policy in Europe*, (pp. 129–60). Berlin: Springer-Verlag.

Alleman, J. H., & Sorce, B. (1997). International settlements: A time for change. In *Proceedings of the Global Networking '97 Conference, Calgary, Canada, 15–18 June 1997* (Vol. 2, pp. 1(8). Amsterdam: ISO Press.

Armstrong, M., Doyle, C., & Vickers, J. (1995, October). *The access pricing problem: A Synthesis.* Paper presented at the PURC/IDEI/CIRANO First Annual Conference, The Transition Towards Competition in Network Industries, Montreal, Canada.

Baumol, W. J., & Sidak, G. (1994). The pricing of inputs sold to competitors. *Yale Journal on Regulation, 11*:171–202.

Carter, M., & Wright, J. (1994). Symbiotic production: The case of telecommunication pricing. *Review of Industrial Organization, 9*: 365–78.

Cave, M., & Donnelly, M. P. (1996). The pricing of international telecommunications services by monopoly operators. *Information Economics and Policy, 8*: 107–23.

Chowdary, T. H. (1997). International accounting rates. *Telecommunications Policy, 21*(1): 77.

Ergas, H., & Patterson, P. (1991). International telecommunications settlement ar-

---

17. Bell Atlantic argues that competing local exchange carriers are not entitled to termination fees for Internet service provider traffic. For Bell Atlantic's position and references to FCC enquiries on this issue, see http://ba.com/policy/positions/1997/Oct/19971029013.html.

rangements: An unsustainable inheritance? *Telecommunications Policy, 15* (1): 29–48.

Galbi, Douglas (1997) "Cross Border Rent Shifting: A Case Study of International Telecommunications," manuscript.

Hakim, Sam Ramsey and Ding Lu (1993) "Monopolistic settlement agreements in international telecommunications," *Information Economics and Policy, 5*: 145–57.

Kahn, A., & Taylor, T. (1994). Comment on Baumol and Sidak's essay. *Yale Journal on Regulation, 11*: 225–50.

Katz, M., Rosston, G., & Anspacher, J. (1995). Interconnecting interoperable systems: The regulators perspective. *Information Infrastructure and Policy, 4*; 327–42.

Kwerel, E. (1984). Promoting competition piecemeal in international telecommunications. (U.S. Federal Communications Commission, Office of Plans and Policy Working Paper Series, No. 13). Washington, DC: Federal Communications Commission .

Kwerel, E. (1987). *Reconciling competition and monopoly in the supply of international communications services: A U.S. perspective.* Paper presented at the Center for Telecommunications and Information Studies' Conference on Asymmetric Deregulation, Paris.

Laffont, J. J., & Tirole, J. (1994). Access pricing and competition. *European Economic Review, 38*: 1673–1710.

Mueller, M. (1996). On the frontier of deregulation: New Zealand telecommunications and the problem of interconnecting competing networks. In D. Gabel & D. Weiman (Eds.), *Opening networks to competition: The regulation and pricing of access* (pp.). New York: Kluwer.

O'Brien, D. P. (1989). *The uniform settlements policy in international telecommunications: A non-cooperative bargaining model of intermediate product third degree price discrimination.* Unpublished doctoral dissertation, Northwestern University, Evanston, Illinois.

Ralph, E. K. (1996). *Regulating an input monopolist with a focus on interconnection in telecommunications.* Unpublished doctoral dissertation, Duke University, Durham, North Carolina.

Stanley, K. B. (1991). Balance of payments, deficits, and subsidies in international communications services: A new challenge to regulation. *Administrative Law Review, 43*(3): 411–38.

Tye, W. (1994). Response to Baumol and Sidak's essay. *Yale Journal on Regulation, 11*: 203–24.

Walker, D. (1995). International accounting rates. *Telecommunications Policy, 20* (4): 239–42.

Yun, K.-L., Choi, H.-W., & Ahn, B.-H. (1997). The accounting revenue division in international telecommunications: Conflicts and inefficiencies. *Information Economics and Policy, 9*, 71–92.

Call-Back and the Proportionate Return Rule: Who Are the Winners and Losers?

Mark Scanlan
Consultant Economist, European Commission

This chapter shows that the operation of the proportionate return rule provides a major, and perhaps the main source of profit for carriers selling call-back minutes. The rule provides for the transfer of incoming International Direct Dial (IDD) minutes and the associated hugely profitable settlement credits, from the other competing carriers, to the carrier selling call-back minutes. In this regard, the rule is not even-handed. The higher the market share of the carrier selling call-back minutes, the less well they do under the rule, so much so that under some circumstances selling call-back minutes at an apparent profit can actually impose losses on the carrier. Counterintuitively, the impact of a call-back service on the financial position of the monopolist in the country where the call-back service is offered will in many cases be positive. The explanation for this result is found by observing cost and revenue streams, and by taking note of the various elasticities and feedback effects wherever there is an increase in traffic. For this operator, the situation is slightly more favorable in a two-country model with call-back than where call-back minutes are terminated in a third country. The policy recommendation suggested in this chapter is to require carriers in *B* (the United States) to make periodic balancing payments that nullify the artificial incentives to sell call-back minutes created by the proportionate return rule.

INTRODUCTION

This chapter analyzes the financial basis for call-back services showing that their viability rests on two institutions: the *international accounting rate system* and the artificially inflated costs this imposes on international public switched telephone network (PSTN) calling, and the *proportionate return* (PR) *rule*. The main findings of this chapter do not agree with perceived wisdom about call-back. The arguments suggest that appeals by operators and officials from many countries to outlaw call-back appear to be based on incorrect presumptions, presumably that call-back is seriously damaging to the operator and/or the country in which the service is offered (call it Country A).[1] Rather, it is demonstrated here that in many instances, even where collection rates in A are three (and possibly more) times the settlement rate with the home country of the call-back service (Country B), the provision of call-back services can increase the profits of the operator(s) in A.

To read this chapter some understanding of the international accounting rate system and the PR rule is required, and thus a brief explanation of the two is set out here.[2]

The *accounting rate system* was developed by administrators in the 1930s when most telecom operators were self-regulating state-owned monopolists. The *settlement rate* almost always splits the accounting rate in two, with a virtual point of interconnection approximately in the middle of the international circuit, hence the term *half circuit*. Thus, the settlement rate was intended to cover the cost from the half circuit to the terminating country, plus the receiving operator's costs to terminate. Original accounting rates may have roughly compensated for the full cost of international calling (i.e., from origin to completion). Each international PSTN minute of telephone traffic (MiTT) registers a settlement liability payable to the terminating country. Periodically, two telecom operators that terminate each other's international calls will settle their account, with the operator placing more outward call minutes to the other making a balancing payment.

A monopoly operator in a country (call it A) unavoidably also has a monopoly for terminating international calls from B. Such a monopolist will want a settlement rate with B that maximizes the rent it receives. If the operator in B is also a monopolist, it will want to do likewise, but prior to knowing the other's profit maximizing price, it is highly unlikely that both operators would arrive at the bargaining table seeking an identical rate.[3] Historically the operator wanting the higher rate (say it is A) has been in the driving seat of this game, and given approximately similar customer characteristics, it is not in B's interest to agree to charge a lesser rate than that being charged by A. Taken in isolation, there is no benefit to B in unilaterally lowering its termination rates for calls from A, as this would result in the capture of rents by foreigners.[4]

Most trade in international telephony traffic is still governed by the accounting rate system, with reciprocal and inflated prices being charged to terminate inter-

---

1. Of the 67 countries that replied to an ITU questionnaire (TSB Circular 217) of 22 April 1996, call-back was illegal in 42 of them.

2. For a more detailed discussion of the international payments system, see Scanlan (1996).

3. For a similar discussion see Cave and Donnelly (1996).

4. When dynamic benefits are taken into account, and there is a multitude of countries trading international traffic, this conclusion may not hold.

national PSTN calls.[5] Traditional trading arrangements are, however, coming under pressure as alternative trading options are increasingly being used.

There are at least two key factors that enable alternative arrangements for trading voice traffic to flourish. First, a limited but increasing number of countries have adopted more liberal regulations that allow for competitors to provide and terminate international telephony services. Many operators have no loyalty to the cartel arrangements that have traditionally existed under the accounting rate system, and their incentives are to trade traffic by alternative (and nontransparent) means if they can do this more profitably. Second, whereas the cost of providing international telephony has collapsed due to technological developments (digital and fiber optic technology), prices have not followed suit. There are thus enormous profits to be made in international telephony.

New trading arrangements for international telephony can be grouped into two, (a) *accounting rate by-pass*, which includes *international simple resale* (ISR; voice over the Internet can be seen as a form of ISR), and (b) those for which accounting rates are paid, but which are able to find some other routing arrangement to provide a cost advantage. These include *least accounting rate routing* (LARR), and call-turnaround, mainly *call-back*.

LARR and accounting rate by-pass, require traffic to be *refiled*. For LARR this can occur when an operator (A) charges discriminatory prices to terminate traffic depending on the country of (declared) origin. It then pays an operator (B) that is being charged a high termination price relative to another operator (C), to come to an arrangement whereby C presents B's traffic to A as if it had come from C.[6]

A proportional return rule is typically adopted by countries that have licensed more than one carrier to supply international services. The rule states that a carrier will receive for termination the same proportion of total incoming traffic from A, as the proportion of total outgoing traffic that it sends to A. The rule is designed to prevent competition between carriers in B to terminate incoming international calls. The need for such a rule arises where foreigners demand rates to terminate that are not cost based, and thus foreign operators have incentives to "whipsaw" between competing carriers in B to secure termination rates that are less than those charged by A to terminate calls from B. The rule prevents competition between B's carriers to terminate incoming international calls, and is aimed at limiting the capture of (net) rents by foreigners.

A *call-back* call never leaves the PSTN. Rather, compared to an international direct dial (IDD) call, the direction of the call is reversed from outgoing to incoming,

---

5. Settlement rates exceed the full economic cost of providing transmission and termination by between 1.5 and 30 times.

6. From this point I use the term *carriers* when referring to the country out of which call-back minutes are provided (almost without exception the United States), but the term operators is used when referring to other countries.

and instead of the originating caller's operator (A) paying a settlement to a foreign operator (B) for termination, A receives a settlement credit.[7] When carriers sell call-back minutes to end users in foreign countries, or to independent call-back services, as is more usual, they are route specific, as the responsible carrier incurs a per-minute settlement rate liability to be paid to A, the receiving operator.

Each call-back minute received by foreign country A carries with it a settlement credit as it is in effect an incoming international call. To the extent that these minutes substitute for outgoing IDD minutes, the operator in A loses its collection revenue on these outgoing calls, but instead receives settlement revenue on the incoming call.

Moreover, if the lower price to the customer induces additional traffic, a significant number of inbound call-back minutes will be new minutes, rather than mere substitution of outbound minutes. On these new minutes the operator in A receives hugely profitable settlement credits without any loss in outbound revenues. The net effect on Country A depends on three elasticities that I discuss in the next section.

For the carrier providing call-back minutes, the outcome depends on whether the called party resides in the same country as the carrier (the United States), or a third country. If it is a third country, settlement rates must be paid to two countries, where only one country requires payment for a call-back call terminating in the home country (B).

This chapter is organized in the following way. The next section provides a discussion of the elasticities that are relevant for the analysis. The section following that analyzes call-back in a two-country world. It begins by presenting a model in the form of stylized international accounts for monopolists in countries A and B, and then compares the changes following the successful introduction in A of a call-back service operated out of B. The next section introduces an own price elasticity effect for call-back. I then introduce the situation where there are two competing carriers in B (B1 and B2) and consider feedback (second period) effects of call-back on IDD calls from Country B to Country A. The next section analyzes the three country case; that is, when a call-back call is terminated in a third country. The conclusion appears in the final section.

THE RELEVANT ELASTICITIES

Assume for the time being a two-country world. Subscribers may undertake non-commercial arbitrage, depending on the call price difference between A and B; that

---

7. An important warning needs to be attached to these results. It seems likely that some services that are marketed as call-back are in fact by-pass or ISR services. These by-pass services include Internet telephony, and double-ended breakout on private networks. The analysis provided by this chapter is not applicable to by-pass services. For one thing, they are not actually call-back services as the direction of the call and the settlement rate liability are not reversed over a PSTN network.

is, the tendency for the party who resides in the country in which calls are cheapest to originate a higher proportion of minutes to the other party.[8] Some call-back minutes (consumed in A) would be in substitution of IDD calls that would have originated in B.[9] This type of arbitrage will almost exclusively relate to social calls, and intuitively, we would not expect the effects to be very significant, an assumption supported by albeit scant empirical evidence. Cheong and Mullins (1991) found no statistically significant impact of international price differences in explaining U.S. international traffic imbalances. Acton and Vogelsang (1992) were unable to detect any feedback effects at all from lower call prices in the United States, on the level of U.S.-bound calls from Western Europe. However, neither of these two studies, or those mentioned below, were intended to analyze the interaction between call-back and IDD calls.

Another effect due to the introduction of a low-priced call-back service in A arises where incoming and outgoing calls are complements, and this is known as *call stimulation*. Call stimulation tends to internalize beneficial call externalities.[10] In terms of the impact on the numbers of outgoing minutes from B to A (where a discounted call-back service begins), arbitrage and stimulation work in opposite directions. Empirical evidence suggests that call stimulation is significant and swamps any subscriber arbitrage effects (where incoming and outgoing calls are substitutes), although the lack of a feedback effect in the Acton and Vogelsang (1992) study is consistent with both effects canceling each other out. For long distance calls in the United States, elasticities of (net) call stimulation as high as 0.75 have been found. Between Canada and the United States the stimulation figures are lower, between 0.24 and 0.40.[11] The stimulation effect produced by new calls originated by call-back subscribers in A on IDD calls from B is examined later. Subscriber arbitrage (roughly joint utility maximizing behavior) is assumed to be insignificant and plays no further part in this analysis.

In the next section I progressively build a model that shows the impact of the two major elasticities underlying call-back: (a) those call-back calls that substitute for IDD calls where the subscriber would have paid a full collection rate to the incumbent operator in A, and (b) those that are wholly new minutes resulting from the lower price of call-back. The analysis implicitly assumes a two-period process. In the first period there are no foreign feedback (stimulation) effects. Indeed, the results of most of this chapter occur in Period 1. In this period the two elasticities—the absolute value of the own price elasticity of demand for call-back ($\mathcal{E}_{cb}$),

8. I assume that commercial arbitrage services are not available to subscribers in Country B.

9. This type of arbitrage is close to joint utility maximization.

10. For an analysis of externalities in telecommunications, see Cave, Milne, and Scanlan (1993).

11. See Larson, Lehman, and Weisman (1990), and Applebe, Snihur, Dineen, Farnes, and Giordano (1988), in Taylor (1994).

and the cross-price elasticity of substitution, call-back for IDD ($\mathcal{E}_{cbIDD}$)—add to 1. This follows from the assumption that putting any second period effects aside, call-back calls will either be in substitution of IDD calls or they will be calls that would not have been made at all except for the lower price of call-back

THE TWO-COUNTRY CASE

A Stylized Example

In a two-country example, A and B, assume a settlement rate of 50 cents,[12] a collection rate in each country of $1 per MiTT, a resource cost for termination from the virtual point of interconnection at the half circuit of 6 cents per MiTT, and that A and B send each other 10 million MiTT. Each telecom operator is making $8.8 million, as shown in Table 5.1. The main points demonstrated by using these assumptions are essentially unaffected by introducing cost or revenue figures that differ between A and B. (A bracketed figure for Country A in Table 5.1 shows the situation where the collection rate is $1.50.)

TABLE 5.1

The Position of Both Carriers A & B Prior to Call-Back

	Country A			Country B		
	Dr	Cr		Dr	Cr	
Collection		$10m ($15m)	*Collection*		$10m	
Revenues		{$1 * 10m MiTT}	*Revenues*		{$1 * 10m MiTT}	
I/C		$5m	*O/G*	$5m		
Settlement		{$0.5 * 10m MiTT}	*Settlement*	{$0.5 * 10m MiTT}		
Payment			*Payment*			
O/G	$5m		*I/C Settle-*		$5m	
Settlement	{$0.5 * 10m MiTT}		*ment Pay-*		{$0.5 * 10m MiTT}	
Payment			*ment*			
Resource	$1.2m		*Resource*	$1.2m		
Cost	2*{6¢ *10m MiTT}		*Cost*	2*{6¢ * 10m MiTT}		
Total	$6.2m	$15 m ($20m)	*Total*	$6.2 m	$15 m	
Net Position		$8.8m ($13.8m)	*Net Position*		$8.8m	

Now assume that the carrier in B begins offering a call-back service in A, and that 1 million call-back minutes are sold to subscribers. It is assumed for ease of exposition that the carrier selling call-back minutes is integrated into the provision of call-back to subscribers in A. The call-back price per minute is not relevant to

---

12. Given U.S. minute weighted average accounting rates over the call-back period, the 50-cent settlement rate is a realistic figure to work with. The following are minute weighted average U.S. accounting rates: 1987 = $1.40; 1990 = $1.28; 1993 = $0.96; 1996 = $0.73.

the analysis that follows. What is important is the assumption that the transfer price between the call-back service and the (owner) carrier is assumed to equal the carrier's full cost of providing those minutes.[13] Thus, taken in isolation, the sale of minutes to the call-back service leaves the carrier selling these minutes in essentially the same financial position as before the sale occurred.[14]

Call-Back as an IDD Substitute

Assume for the moment a worst case scenario, that all the call-back operator's traffic is in substitution of IDD calls originated on the incumbent operator A's network. In this case A would lose collection revenues of $1 million (1 million MiTT × $1). However, as A would now not have to pay the outgoing settlement rate on these minutes, the cost of losing these IDD minutes is $.5 million {$1 million – $.5 million}. The 1 million MiTT are now no longer directed out of the country, but have been turned around, and the carrier in country B that provides the outgoing call minutes for the call-back service must now pay the per-minute settlement rate of $0.50, a total of $0.5 million to the operator in A.

TABLE 5.2
A & B's Position Following Provision of
a Purely Substitute Call-Back Service in A

	Country A			Country B	
	Dr	Cr	.	Dr	Cr
Collection Revenues		$9m ($13.5)^θ {9m MiTT * $1}	Collection Revenues		$10m {10m MiTT * $1}
I/C Settlement Payment		$5.5m {$0.5*10m +$0.5 * 1m MiTT}	O/G Settlement Payment	$5.5 m {$0.5*10m + $0.5 * 1m MiTT}	
O/G Settlement Payment	$4.5m {9m MiTT * $0.5}		I/C Settlement Payment		$4.5m {$0.5*9m MiTT}
Resource Cost	$1.2m {6¢*11mMiTT}+{6¢ *0.9m MiTT}		Resource Cost	$1.2m {6¢*11mMiTT}+{6 ¢*0.9m MiTT}	
Receipt from CB			Receipt from CB		$0.56m
Total	$5.7m	$14.5m ($19)	Total	$6.7m	$15.06m
Net Position		$8.8m ($13.3)	Net Position		$8.36m

Notes: θ The bracketed figures apply where the collection rate in A is $1.50.

On the assumption that the resource costs of sending and receiving calls is the same, there is no change in either operator's resource cost due to the turnaround

---

13. This will include a return on capital employed. The simplest cost calculation involves capacity costs spread over (forecasted) minutes.

14. As income effects will be tiny, they have not been included in the analysis.

in the direction of this traffic. The conclusion, given the worst case scenario that call-back minutes only substitute for A's IDD minutes, is that at a $1 IDD collection rate in A the operator in A is no worse off than before the call-back service began operating, remaining at $8.8 million. On the other hand the net position of the carrier in B, which is providing the call-back service in A, has deteriorated from $8.8 million, to $8.36 million, and its average per-minute profitability of B has also declined (Table 5.2).

Combining an Own Price Elasticity Effect

It was assumed earlier that the minutes of traffic provided by call-back were simply in replacement of IDD calls from A to B. But as well as a substitution effect between IDD and call-back, there will also be an increase in the total number of international call minutes initiated by subscribers in A due to the cheaper price of the call-back service. These are calls that would not have been made at the price (collection rate) of the incumbent IDD provider in B.

TABLE 5.3
A & B's Position When B Also Provides a Call-Back Service in Country A, Which Combines Both New and IDD Substitute Calls

| | Country A | | | Country B | |
	Dr	Cr		Dr	Cr
Collection Revenues		$9.5m (14.25) {9.5m MiTT * $1}	Collection Revenues		$10m {10m MiTT * $1}
I/C Settlement Payment		$5.5m {$0.5*10m + $0.5* 0.5m MiTT} + 0.5m MiTT(new calls)* $0.5	O/G Settlement Payment	$5.5 m	{11m MiTT* $0.5}
O/G Settlement Payment	$4.75m {9.5m $0.5m}	MiTT*	I/C Settlement Payment		$4.75m {9.5m MiTT*$0.5m}
Resource Cost	$1.23m {6¢ *20.5m MiTT}		Resource Cost	$1.23m {$6¢ *20.5m MiTT}	
Receipt from CB			Receipt from CB		$0.56m
Total	$5.98m	$15m (19.75)	Total	$6.73m	$15.31m
Net Position		$9.02m (13.77)	Net Position		$8.58m

Estimates of the price elasticity of demand for international calls start at negative 0.6 and get absolutely larger.[15] The explicit assumption with this and similar figures is that they are obtained by observing price changes involving an IDD service. The Country A price elasticity effect explained by the lower price of call-

15. See the review of empirical work on telecommunications-related elasticities in Taylor (1994).

back is the own price elasticity of demand for call-back, not the price elasticity for IDD calls. There is no empirical evidence known to this author of the price elasticity of call-back, or the cross-price elasticity of substitution, call-back for IDD.

FIGURE 5.1

The impact of call-back on country/operator revenue in a 2-country model.

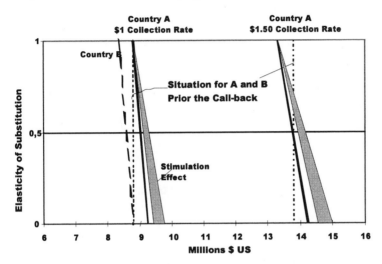

An extension of the example just outlined to include an own price elasticity effect is shown in Table 5.3, on the assumption that of the 1 million minutes of call-back consumed in Country *A*, half a million of them substitute for IDD calls that would have been billed by the incumbent operator, and the other half a million minutes is new traffic. I later demonstrate the sensitivity of this assumption.

Note that it is assumed that 1 million minutes of call-back are sold whatever the price elasticity of call-back is. Thus, the incumbent operator in *A* loses 0.5 million of international MiTT, or $0.5 million ($1 × 0.5 million) in call revenue, less the $0.25 million (the settlement cost for this traffic payable by *A* to *B*) equals $0.25 million. On the other hand, *A* receives an additional 500,000 incoming minutes of completely new traffic from *B*, and 500,000 incoming minutes that have substituted for IDD calls from *A* to *B*. Together they add up to $0.5 million of additional incoming settlement payments.

Compared with the net positions of *A* and *B* from Table 5.2, *A*'s and *B*'s positions have improved from $8.8 million to $9.02 million for *A*, and for *B*, from $8.36 million to $8.58 million. Compared to the no call-back scenario in Table 5.1, *A* has moved from $8.8 million to $9.02 million, but *B* is still worse off.

For the operator in *B*—the call-back sponsor—the main drawback in providing a call-back service is the substitution of IDD calls from *A* to *B* (and the loss of settlement income) due to the call-back alternative. The higher is the ratio of IDD substitute traffic over newly generated traffic due to call-back, the higher the call-back collection rate will have to be before it is to B's benefit to sponsor a call-back service. This situation is shown graphically in Figure 5.1, under the assumption that the settlement rate is 50 cents. Two situations are shown: a $1 and $1.50 IDD collection rate in A.

The impact of call-back on monopolists A and B depends on the ratio of the elasticity of substitution, call-back for IDD (ε_{cbIDD}), over ε_{cbIDD} plus the absolute value of the own price elasticity of demand for call-back (ε_{cb}). This ratio appears on the vertical axis. Although we do not in fact know the values of the various elasticities involved, it nevertheless seems safe to assume in a two-country model that with a collection rate up to about three times the settlement rate, a call-back service will not weaken Operator A's financial position. Note that Country B is worse off whenever $\varepsilon_{cbIDD} > 0$.

When B Has Two Competing Carriers and a Proportionate Return Rule

Assume now that B has two licensed carriers ($B1$ and $B2$) that compete to sell international calls to subscribers in B, but a PR rule is in operation. Let us say Operator $B1$ has 90% of the market for IDD calls to A, and $B2$ has a 10% share, and that traffic between Countries A and B is in balance. Prior to call-back, Operator A's position remains as in Table 5.1. The position of the carriers $B1$ and $B2$ add to the same net position, $B1$ and $B2$ with 9 million and 1 million MiTT, respectively.

Assume now that $B2$ sells 1 million voice minutes at cost to a service that offers call-back in A (shown in Table 5.4). As earlier, it is assumed that half of the call-back traffic substitutes for calls that would have used A's IDD service, and half the 1 million MiTT are assumed to be new calls due to the lower price of call-back.

For Country B, the loss of incoming call-minutes and subsequent settlement revenues due to the turnaround of 0.5 million MiTT of IDD traffic into call-back impacts on $B1$ and $B2$ disproportionately, in accordance with the different proportions that each operator has of the outgoing MiTTs to A. $B2$'s minutes to A increase by 1 million, and although the outgoing IDD minutes from A have declined (from 10 million MiTT to 9.5 million MiTT), $B2$ now gets 2/11 of 9.5 million incoming minutes, instead of 1/10 of 10 million MiTT it had received before providing the call-back minutes. $B1$ goes from terminating 9 million MiTT to terminating 7.773 million MiTT (9/11 – 9.5 million MiTT), a loss of 1.227 million incoming minutes, equal to $0.614 million in settlement revenues. For relatively small competing operators such as $B2$, selling call-back minutes provides a way of transferring hugely profitable incoming minutes from $B1$ to itself. Indeed, for small market-share carriers, their main interest is not likely to be in trading call-back minutes per se, but in using the PR rule to transfer incoming minutes from the other carrier(s) to themselves.

TABLE 5.4

The Position of Carriers A, B_1, & B_2, Following the Provision of Call-Back by B_2

	Operator B_1		Operator B_2	
	Dr	Cr	Dr	Cr
Collection Revenues		$9m {9m MiTT * $1}		$1m {1m MiTT * $1}
I/C Settlement Payment		$3.886m {9/11 * 9.5m MiTT} *$0.5		$0.864m {2/11 * 9.5m MiTT} * $0.5
O/G Settlement Payment	$4.5m {9m MiTT * $0.5}		$1m$^\Phi$ {1m MiTT * $0.5} + {1m MiTT * $0.5}cb	
Resource Cost	$1.006m$^\delta$ {9m MiTT * 6¢} + {9/11*9.5mMiTT*6¢}		$0.224m {2mMiTT*6¢}(o/g)+ {2/11*9.5mMiTT*6¢}	
Receipt from Callback for MiTTs				$0.56m
Total	$5.506m	$12.886 m	$1.224 m	$2.424 m
Net Position		$7.38m Net Position		$1.2m

Net Position of Country B $8.58 m

Notes: $^\Phi$Operator B2 must also pay for the settlement cost for outgoing minutes provided by the call-back service. The corresponding credit from the call-back service appears in the row below.

$^\delta$The first bracket provides the cost calculation for each operator's outgoing minutes. The second bracket shows a similar calculation for incoming minutes.

Figure 5.2 shows B_2's marginal profit on call-back minutes as a function of market share, under different elasticity assumptions. The model used to generate this figure is implicit in Table 5.4. It is important to recall that B_2's transfer of minutes to a call-back service is at a price that just covers its cost, where the call-back service pays the carrier for the settlement rate out-payment and the resource cost associated with each call-back minute. The solid lines all begin at $0.44, because with no market share at all, the first call-back minute sold enables the new carrier to capture virtually a full incoming minute (when traffic between A and B is in balance) and the associated settlement rate credits from the incumbent carrier, less the associated resource cost. If this were 10 cents per minute rather than the 6 cents assumed, then the first call-back minute sold would provide a net benefit of $0.40, as shown by the dotted line, which is drawn for $\mathcal{E}_{cbIDD} / \mathcal{E}_{cbIDD} + |\mathcal{E}_{cb}| = 0.5$. The amounts indicated at specific market shares show the carrier's net financial reward from selling an additional call-back minute at a price intended to cover the direct costs incurred. For these reasons, Figure 5.2 has been referred to as a *discrimination function*.

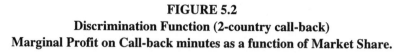

FIGURE 5.2
Discrimination Function (2-country call-back)
Marginal Profit on Call-back minutes as a function of Market Share.

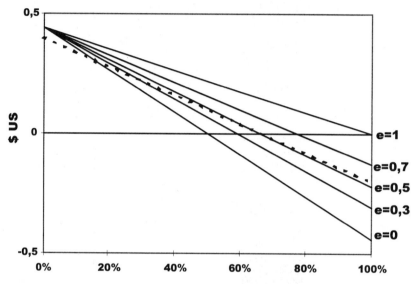

. In the model of two-country call-back, the PR rule does not preserve the alignment between national producer surplus and the producer surplus of individual producer interests, which it needs to do if it is to work sensibly. Where there is more than one carrier in B and the PR rule operates, profit maximizing behavior by the small-market-share operator reduces the producer surplus secured by B's industry.

Moreover, the minimum profitable call-back price offered in Country A will be significantly lower than the break-even call-back price needed by Country B to maintain its average per-minute profitability on its international sector. Thus, $B2$ could sell call-back minutes at less than the settlement rate that must be paid on each route minute. In a similar vein, independent call-back companies may be able to force the price of call-back minutes below the settlement rate payable on those minutes. In our example, this surplus amounts to the change in the net position of $B2$ between the pre-call-back situation and its position in Table 5.4 ($1.20 million – $0.88 million), which is $0.32 million. $B2$'s break-even price for these call-back minutes sold to a call-back service is only 24 cents per outgoing minute to Country A, although the settlement rate payable to A is 50 cents per minute.

Call Stimulation

The advent of a cut price call-back service in A will give rise to feedback effects on the number of IDD calls from B.[16] In order to view the impact of call-back, it is assumed in this chapter that commercial arbitrage (call-back) is only available to *A*'s subscribers.

Where incoming and outgoing calls are complements, each additional call completed has some positive probability of eliciting a return call by the receiving party, a call stimulation effect. This effect shifts the impact of call-back in a beneficial direction for *A*, and also lessens the detrimental impact of the call-back service on the combined financial positions of *B*'s carriers. The scant empirical evidence that exists suggests that newly generated traffic will stimulate new incoming IDD calls from *B* to *A*, in the proportion of somewhere between 2 to 5 incoming (into *B*) minutes for every 10 new minutes generated by call-back (see earlier discussion). If 50% of call-back minutes are newly generated (i.e., had there been no call-back service they would not have been made) then for every 10 call-back minutes terminating in *B*, there would be an additional 1 to 2.5 minutes of incoming IDD traffic into *A*.

The implication of call stimulation for the combined accounts of *B1* and *B2* (Table 5.4; shown in Table 5.3 without the inclusion of a (second period) call stimulation effect), is to expand collection revenues by $0.1 to $0.25 million, but with corresponding increases in outgoing settlement payments and resource costs. The impact of call stimulation on the two-country model can be seen in Figure 5.1.

THE THREE-COUNTRY CASE

The provision of a call-back call often involves three countries: *A*, where the call-back service is offered; *B*, the home country of the carrier providing the outgoing call minutes to the call-back service; and *C*, where the call-back call is terminated. In almost every case, a legal call-back service involving three countries requires two outgoing international public switched calls from *B*, one to the originating caller in *A*, and the other to terminate in *C*.[17] Assume each country sends and receives 20 million MiTT. As before, assume *B1* has 90% of *B*'s outgoing international minutes to each destination, and *B2* has 10%. While countries *B*'s IDD collection rate is given at $1 per minute (as in the two-country case), for *A* and *C* the rate is assumed as $1.50 so that there is sufficient margin for a call-back service to operate. The cost to *B2* of the outgoing minutes it supplies to complete a call-back call from *A* to *C* is $1.08 per call-back minute.[18]

Stylized accounts are shown in Table 5.5, where the call-back minutes are for termination in *C*. Call-back has resulted in Operator *A* losing $0.75 million in IDD collection revenues, but receiving an additional $0.5 million in settlement from the call-back service. Moreover, with the loss of outgoing minutes due to the substitution of IDD by call-back, *A*'s settlement liability has been reduced by $0.25 million.

---

16. Note that call-back does not alter the direction of call stimulation.

17. Exceptions might occur when ISR is permitted between Countries B and C.

18. {[$0.50 (settlement rate)] * 2} + {6¢ (termination in C) + 2¢ (transit through B)}.

TABLE 5.5
The International Accounts for Carriers A, B_1, B_2, and C, After the Inclusion of a Call-Back Service Provided by B_2 in A for Calls to C

	Carrier A			Carrier C	
	Dr	Cr		Dr	Cr
Collection Revenues		$29.25m {19.5m MiTT * $2}			$30m {20m MiTT * $2}
I/C Settlement Payment		$10.5m {10m MiTT * $0.5}+ {11m MiTT * $0.5}			$10.25m {9.5m MiTT * $0.5}+ {11m MiTT * $0.5}
O/G Settlement Payment	$9.75m {10m MiTT * $0.5}+ {9.5m MiTT * $0.5}↑			$10m {10m MiTT * $0.5}+ {10m MiTT * $0.5}+	
Resource Cost	$2.43m {19.5m MiTT*6¢pm} {20m MiTT*6¢pm} {1m MiTT * 6¢pm}†			$2.43m {19.5mMiTT * 6¢pm} {20m MiTT * 6¢pm} {1m MiTT * 6¢pm}	
Total	$12.18	$39.75m		$12.43m	$40.25m
Net Position		$27.57m	Net Position		$27.82m

	Carrier B1			Carrier B2	
	Dr	Cr		Dr	Cr
Collection Revenues		$18m ($18)ᵅ {9m MiTT* $1} {9m MiTT * $1}			$2m ($2m) {1m MiTT* $1} {1m MiTT * $1}
I/C Settlement Payment	2(9/10*10*.5)= $9m	$8.182m ($9m) {9/11 * 10m MiTT * $0.5} + {9/11 * 10m MiTT * $0.5}		2*(1/10*10*.5)= $1m	$1.818 ($1m) {2/11 * 10m MiTT * $0.5} + {2/11 * 10m MiTT * $0.5}
O/G Settlement Payment	$9m ($9m) {9m MiTT * $0.5}+ {9m MiTT * $0.5}↑			$2m ($1m) {2m MiTT*$0.5}+ {2m MiTT*$0.5}←	
Resource Cost	$2.062m ($2.16m) [(2{9/11 * 10mMiTT)) +{18m MiTT}]* 6¢pm	[2*{9/10*10mMiTT} +18mMiTT]* 6¢pm =$2.16m		$0.418m ($0.24m) [2[{2/11*10m MiTT} + {3m MiTT] * 6¢pm +1m MiTT * 2¢pm†	[2*{1/10*10mMiTT} +2mMiTT]* 6¢pm =$0.24m
Receipt for Cb MiTTs					$1.08m
Total	$11.062m ($11.16)	$26.182m ($27m)		$2.4182m ($1.24m)	$4.898m ($3m)
Net Position		$15.12m ($15.84m)	Net Position		$2.480m ($1.76m)

Notes: ↑ A now sends 9.5 million IDD minutes to C.
†This is the resource cost to terminate the incoming call-back minutes from B1.
←A settlement debit for the 1 million minutes of call-back traffic is incurred on both outgoing legs.
ᵅThe underlined figures in brackets are those that apply to B1 and B2 prior to a call-back service being operated.

The profit of carrier B2 has improved from $1.76 million to $2.48 million. At the level of B2's consolidated accounts (i.e., including the call-back service), the break-even price of call-back is 36 cents per minute,[19] leaving a loss on call-back

---

19. B2's revenues ($3.818 million) – B2's direct costs ($2.418 million) = $1.4 million, plus $0.36 million = $1.76 million.

appearing in B_2's accounts of $0.72 million, which is the amount of the transfer in settlement credits from B_1.

FIGURE 5.3

The impact of call-back on operator/carrier revenue in a 3-country model.

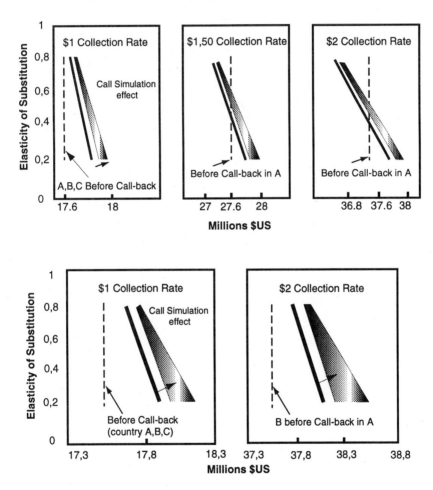

Thus, were B_2 not to own the call-back operator, B_2 could still price outgoing call-back minutes from B to A, and B to C, together, as low as 36 cents per minute, even though it had to pay two times $0.50 per minute in settlement to operators A and C. The difference (32 cents per route, excluding resource costs) is what it receives from $B1$, through the operation of the PR rule. Clearly, by selling at below settlement, $B2$ is pursuing a rational policy.

Figure 5.3 shows the impact of call-back in a three-country model, with the elasticity appearing on the vertical axis being the same one that appears in Figure 5.1, $\mathcal{E}_{cbIDD}/\mathcal{E}_{cbIDD}+|\mathcal{E}_{cb}|$. The horizontal axis indicates the net financial positions of the carriers and operators. The various scenarios shown are for different IDD collection rates. For Operator A, even if the elasticity $\mathcal{E}_{cbIDD}/\mathcal{E}_{cbIDD}+|\mathcal{E}_{cb}| = 0.8$, call-back to a third country via $B2$ is always to its benefit given that its IDD collection rate is not more than approximately 220% of the settlement rate. Indeed, as can be seen from Figure 5.3, where A's collection rate is four times the settlement rate, and where the elasticity of substitution, call-back for IDD calls ($\mathcal{E}_{cbIDD}$), is less than about 0.4 to 0.3, the provision of a call-back service in Country A will still be to Operator A's advantage.

Operator C always benefits from a genuine call-back service offered in A. This is because whether the calling party in A uses IDD or call-back, C still receives a settlement rate credit. As the lower price of call-back results in more calls being made, C gets more settlement income.

FIGURE 5.4
Discrimination Function (3-country call-back).
The marginal profit of call-back minutes as a function of market share.

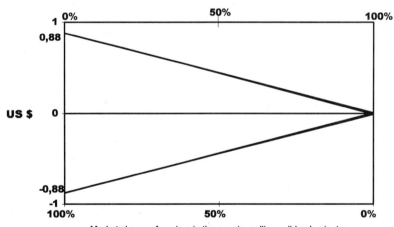

Market shares of carriers in the country selling call-back minutes

Assuming a fixed number of call-back minutes are sold by $B2$ at a price that just covers its long-run average incremental cost, the collection rate assumed in A has no bearing on $B2$. The benefit that $B2$ enjoys from the sale of these minutes is entirely to the cost of $B1$. It is simply a transfer of settlement credits through the operation of the PR rule to the carrier selling the call-back minutes, from the other carrier(s). Moreover, the size of the transfer $B2$ receives depends on its market share.

Figure 5.4 plots the marginal profit per call-back minute sold in the three-country model, as a function of market share. The upper function, indicating positive dollar values, is for the carrier selling call-back minutes. The lower function, indicating negative dollar values, represents the loss of incoming settlement credits net of capacity cost savings, with each minute of call-back sold by the other operator(s). Note that there is no loss of surplus from Country B as there was in the two-country model due to the substitution of call-back for incoming IDD calls from A.

As opposed to the two-country call-back scenario, in the three-country case, even a high-market-share carrier makes money by selling call-back minutes at cost; it simply makes less money the greater its market share. Where a carrier has no market share at all, its first call-back minute sold will return it two settlement credits, less the relevant resource costs. If the settlement were 40 rather than 50 cents, the first call-back minute would return the carrier 68 cents, rather than the 88 cents shown.

CONCLUSION

For those countries and operators in which call-back services are on offer (Country A) and that are worried about the financial implications for the national operator(s), it seems clear there is rather less to worry about than they think.[20] In a two-country world, they can be relatively sure that where IDD collection rates are not more than about three times the settlement rate, call-back will contribute positively to the profitability of Operator A. Indeed, if the elasticities are favorable, call-back will not undermine the financial position of the operator (A) at ratios even higher than this.

In a three-country case (i.e., where call-back calls are not terminated in the home country of the call-back service provider), the situation for A is only slightly less favorable than in a two-country world. If the preponderance of call-back calls are in substitution of IDD calls from A, call-back services that terminate in a third country will negatively impact on Operator A's financial position if its IDD collection rates are more than about 2.2 times the settlement rate. On the other hand, if the elasticities are especially favorable, call-back will add to A's financial position,

---

20. I have not attempted to analyze consumer interests, externalities, or the broader economic welfare of countries in this chapter.

and this is possible even for IDD rates as high as five or more times the settlement rate, as shown in Figure 5.3.

The implication for the country (and its carriers) that originates the outgoing call-back minutes (Country B, the United States) is not so favorable. In a two-country world, selling call-back minutes at a normal (or competitive) rate of return on capital lowers the average per-minute and absolute profitability of the international tele-communications sector in B. Where there is competition between carriers in B, the loss of surplus is felt by the large-market-share carrier, the gains by the small-market-share carrier(s) ($B2$) being on average insufficient to counterbalance $B1$'s loss.

In a three-country call-back model, $B1$'s and $B2$'s consolidated accounts are un-affected by one of them selling call-back minutes at a price that just covers its re-source cost. However, for the large-market-share carrier ($B1$), the implications of $B2$ selling call-back minutes are even more unfavorable than under the two-coun-try model, with incoming settlement credits on both routes (to A and simulta-neously to C) being transferred from $B1$ to $B2$ through the operation of the PR rule.

If settlement rates remain significantly above cost, the policy recommenda-tion suggested by this chapter is to require carriers in B to make periodic bal-ancing payments that nullify the artificial incentives created by the PR rule to sell call-back minutes. This would require a calculation involving turned around minutes and route-specific resource costs and settlement rates to work out compensatory payments.

The question arises as to why so many operators and policymakers are vehemently opposed to the provision of call-back in their own countries given the results of this chapter (see footnote 1). The answer may well relate to the complexity of internation-al telecommunications trade. It appears that a good deal of study is required in order to obtain a working knowledge of its peculiarities.[21]

ACKNOWLEDGMENTS

The views expressed in this chapter are not necessarily shared by the European Commission. The author would like to thank Greg Rosston, Jeff MacKie-Mason, David Waterman, and two anonymous referees for their very helpful comments addressing an earlier draft of this chapter. The usual disclaimer applies.

---

21. Having spoken in early 1997 with executives from AT&T, it was clear that they had an instinctive understanding that the PR rule played an important role that was not always to their advantage. But had AT&T fully comprehended the discriminatory nature of the rule, I would have expected them to have made a forceful case to the FCC for regulatory reform such as I describe in the conclusion. The United Kingdom has also operated a PR rule, but Mercury does not seem to have ever entered the business of exploiting the rule by selling call-back minutes.

REFERENCES

Acton, J. P., & Vogelsang, I. (1992). Telephone demand over the Atlantic: Evidence from country-pair data. *The Journal of Industrial Economics, Vol. XL,* 305–332.

Applebe, T. W., Snihur, N. A., Dineen, C., Farnes, D., & Giordano, R. (1988). Point-to-point modeling: An application to Canada–Canada and Canada–U.S. long distance calling. *Information Economics and Policy, 4,* 311–331.

Cave, M., & Donnelly, M. P. (1996). The pricing of international telecommunications services by monopoly operators. *Information Economics and Policy, 8,* 107–123.

Cave, M., Milne, C., & Scanlan, M. (1993). *Meeting universal service obligations in a competitive telecommunications sector.* Brussels-Luxembourg: ECSC-EEC.

Cheong, K., & Mullins, M. (1991, April). International telephone service imbalances *Telecommunications Policy,* 107–118.

Larson, A. C., Lehman, D. E., & Weisman, D. L. (1990). A general theory of point-to-point long distance demand. In A. de Fontenay, M. H. Shugard & D. S. Sibley (Eds.), *Telecommunications demand modeling,* (pp. 299–318). Amsterdam: North-Holland.

Scanlan, M. H. (1996). Why is the international accounting rate system in terminal decline, and what might be the consequences? *Telecommunications Policy, 20,* 739–753.

Taylor, L. (1994). *Telecommunications demand in theory and practice.* Boston: Kluwer.

III

THE MEDIA

6

Video Competition and the Public Interest Debate

Howard A. Shelanski
University of California at Berkeley, School of Law

Consumer welfare has been a central theme in the public interest regulation of broadcasters. The Federal Communications Commission (FCC) has exercised its public interest authority to ensure that licensees heed audience demands and offer viewers a broad range of programming. The debate surrounding broadcast regulation has thus consistently focused on economic questions: Do failures in the video market warrant remedial programming guidelines? Does license scarcity justify regulatory demands in return for licensees' valuable spectrum rights? New rules have been justified on the ground that market failures persist. Regulation has been repealed when the market is perceived to serve consumers well. Program diversity has been a crucial measure in the FCC's assessment of video market performance.

Although radical changes have occurred in the video market over the last 15 years, the public interest debate continues today on often the same terms as in the past. The White House has charged an advisory committee with deciding what public benefits should be asked of broadcasters in return for free digital spectrum. Some policymakers find that market failures and spectrum scarcity make it "reasonable to put all media under some obligation to serve the public interest" through "quantified" programming obligations (Hundt, 1996a, 1996b). Others contend that increased competition moots the need for further regulation (Quello, 1995). Economic language thus remains central to public interest regulation: An imperfect market gives licensees insufficient incentive to supply some products consumers want, and public interest rules are designed to remedy the gap.

There is, however, good reason to question whether the economic terms of the past remain useful to today's public interest debate. On one hand, economic justifications for programming rules have lost force. Changes in the amount and structure of video competition have made market failures less likely and regulatory remedies for such failures less feasible with respect to conventional measures of consumer welfare. This suggests that regulation designed merely to enhance program diversity is unnecessary and perhaps harmful, an argument that has by now often been made, but which gains strength in light of market developments.

On the other hand, economic goals have become less relevant to public interest regulation. Regulatory priorities have partly shifted from the diversity-oriented policies of the past to narrower programming goals that may have little to do with the kinds of market failure or consumer welfare considerations traditionally underlying the public interest debate. Policy efforts to produce more political coverage or children's programming, for example, may have more to do with normative social judgments than with correcting market inefficiencies or maximizing current consumer satisfaction. This suggests that perhaps the more important public interest rules today are not usefully discussed in terms of correcting for, or being mooted by, the market. Accordingly, a further shift in the public interest debate away from its traditional economic terms has become necessary to determine the plausibility and wisdom of current video program regulation.

In this chapter, I first describe the historical cycle of public interest regulation as it has applied to program diversity and review the market failure justification for programming rules. I then examine how recent developments in the video market affect the goals and justifications of program regulation and discuss the consequences for the current public interest debate.

PUBLIC INTEREST REGULATION OF PROGRAMMING

Regulators' concern with the programming market is as old as commercial broadcasting itself. The Radio Act of 1927 expressly required licensees to ensure that programming was in keeping with the "public interest, convenience and necessity" and regulators warned licensees against using their freely bestowed spectrum rights to air programs that were "uninteresting," "distasteful," or not "consistent with the public service expected of the station" (Federal Radio Commission, 1928, 1929). In 1931 the Federal Radio Commission removed the most popular show in the country from the airwaves, partly because the show was making fraudulent medical claims that could harm the public. (*KFKB Broadcasting v. Federal Radio Commission*, 1931). After congress passed the Communications Act of 1934, which incorporated the 1927 Act's "public interest, convenience and necessity" language, the newly created FCC imposed sanctions on stations for airing personalized astrological forecasts and coded horse-racing results intelligible to listeners who had mailed in a fee, (*Scroggin & Co. Bank*, 1935; *Bremer Broadcasting Co.*, 1935). The FCC ruled that such programming "could not reasonably be said to have any general interest for the public" (*Scroggin & Co. Bank*, 1935). Early regulators thus used their public interest authority defensively, to ensure that "harmful" or "uninteresting" programming did not stay on the air to the public's detriment. Their direct concern was not with what stations were failing to program, but with what they were actually broadcasting.

The Supreme Court initially interpreted the FCC's public interest authority under the 1934 Act to be quite limited. In *FCC v. Sanders Brothers Radio* (1939), the Court found that under the 1934 Act "the Commission is given no supervisory control of the programs, of business management, or of policy" of licensees (p. 475). With respect to such matters, the Court found that "Congress has not, in its regulatory scheme, abandoned the principle of free competition" (p. 474). Just 3 years later, however, the Supreme Court ruled in *NBC v. United States* (1942) that "the Act does not restrict the Commission merely to supervision of the [broadcast] traffic. It puts upon the Commission the burden of determining the composition of that traffic" (pp. 215–16).

The Court's revised conclusion was motivated by scarcity: "The facilities of radio are not large enough to accommodate all who wish to use them. Methods must be devised for choosing from among the many who apply" (p.216). In *Ashbacker Radio v. FCC* (1945), the Court ruled that those methods must include comparative hearings on competing license applications. In an initial effort to supply evaluative criteria, the FCC in 1946 issued its *Blue Book* report requiring licensees applying for renewal to prove they had aired programming responsive to local, minority, and noncommercial interests (FCC, 1946).

Although the 1946 guidelines themselves were not specifically enforced, more expansive content regulations on broadcasters eventually were. Revisiting its programming rules in the wake of the television quiz show scandals, the FCC in 1960 listed 14 "major elements" necessary for broadcasters to meet their public interest duties (FCC, 1960).[1] The FCC's internal enforcement rule for the 1960 categories was to withhold action on license applications that did not provide, inter alia, plans for at least 5% "local programming," 10% prime-time "sustaining programming," and at least some religious, agricultural, educational, news, and discussion programming (FCC, 1983)

Competitive license application proceedings, already extensive and comparative affairs for the FCC in the wake of *NBC* and *Ashbacker*, became increasingly unwieldy when the FCC had to evaluate how well competing program proposals might satisfy the 14 criteria of the 1960 policy. The FCC's effort in 1965 (FCC, 1965) to streamline programming criteria failed to lift an applicant's burden to justify its case "on the basis of program offering...and any other matters the parties asked the FCC to consider as pertaining to licensee fitness" (*Greater Boston Tele-*

---

1. The 14 categories were opportunity for local self-expression, development and use of local talent, programs for children, religious programs, educational programs, public affairs programs, editorialization by licensees, political broadcasts, agricultural programs, news programs, weather and market reports, sports programs, service to minority groups, and entertainment programs.

vision Corp. v. FCC, 1970). The courts rebuffed the FCC's simultaneous effort to exempt many renewal applications from full-fledged hearings (*Office of Communication of United Church of Christ v. FCC*, 1966). The FCC received 447 petitions challenging 936 broadcast licenses from 1970 to 1977 alone (Teeter & Le Duc, 1992).

In 1976 the FCC decided the broadcasting market was working sufficiently well to warrant deregulating radio programming and leaving content entirely in licensees' discretion (FCC, 1976a). The FCC had periodically expressed hesitation to act as a "national arbiter of taste" (Palmetto Broadcasting, 1962), nonetheless finding the goal of program diversity to merit "preservation of a format [that] would otherwise disappear" when the FCC was so petitioned (*Citizens Committee to Save WEFM* , 1974). The FCC's 1976 order on program formats concluded that the market alone should judge a licensee's ability to "survive or succumb according to his ability to make his programs attractive to the public" (FCC, 1976a, p. 861). The Supreme Court eventually upheld the FCC's policy, finding sufficient evidence that it would "further the interests of the listening public as a whole [to rely] on market forces to promote diversity in radio entertainment formats and to satisfy the preferences of radio listeners" (*FCC v. WNCN Listeners Guild*, 1981).

Having gotten out of the business of regulating program format, the FCC still had in place minimum local and informational programming requirements that transcended basic format decisions; that is, rules that would apply whether the station broadcast rock, opera, jazz, or sports. In 1976 the FCC had set minimum, quantified levels of such programming that would allow a radio or television broadcaster to escape full review of its renewal application (FCC, 1976b). Under those rules the FCC delegated renewal authority to the Broadcast Bureau in every case where AM or FM radio licensees respectively proposed at least 8% and 6% nonentertainment programming and where television broadcasters proposed at least 5% each of local and information programming. The FCC decided to jettison these programming rules for radio licensees and, in 1984, to remove programming requirements for television broadcasters, too.

In a lengthy order, the FCC concluded that elimination of television programming guidelines was justified because broadcasters had sufficient market incentives to produce the programming desired by viewers (FCC, 1984). The FCC noted that production of "local" and "informational" programming had not declined significantly on UHF stations previously exempted from the FCC's public interest guidelines. The FCC also found that the promise of new technologies coupled with the expansion of cable and of broadcasters themselves would, through competition, "further ensure the presentation of sufficient amounts of [local and informational] programming" (¶ 21). The FCC thus concluded that "licensees should be given [the] flexibility to…alter the mix of their programming consistent with market demand" (¶ 23).

In keeping with its deregulation of broadcasting, the FCC declined to allow programming considerations to affect licensing of new video technologies. Even before it repealed the broadcasters' program guidelines, the FCC had refused to extend those requirements or other public interest rules to direct broadcast satellite (DBS) operators. Conventional broadcasters opposed that decision, which the FCC successfully defended in court. The FCC noted that among the likely benefits of DBS was programming better suited to particular viewers' tastes, spurred in part by increased competition for advertising revenues that would result from entry by DBS into the video market (FCC, 1982a, 1982b). The FCC later removed public interest obligations from other developing technologies like Multichannel Multipoint Distribution Systems (MMDS), a "wireless cable" service, reversing decisions of the 1970s that had subjected such licensees to public interest regulation. (FCC, 1986, 1987).

The market mechanism and its perceived effects on program diversity thus played a central role in the historical cycle of public interest regulation just sketched. Although broadcasting policy certainly entailed social goals, there was a distinctly economic cast to the FCC's justifications for programming rules: Regulation was presented as a remedy for the effects of scarcity and for the market's failure to produce viewers' desired mix of programming; deregulation occurred where the FCC found market competition superior to intervention.

CURRENT REGULATION

Current regulatory initiatives take various forms. In some cases the FCC has reserved the right to impose future, unspecified "public interest" requirements on licensees. The FCC's local multipoint distribution service (LMDS) order is an example (FCC, 1997a). In other cases the FCC's rules are much more precise. Regulations implementing the Children's Television Act (1990) are probably the best examples. Free time for political candidates has also become a focus of current regulatory debate, as have quantified set-asides for noncommercial educational and informational programming.

Participants in the current regulatory debate often phrase the goals and justifications for new programming rules in economic terms similar to those that guided the past debate. For example, echoing Justice Frankfurter's *NBC* opinion, former FCC Chairman Hundt (1996b) said regulation is necessary because "there just aren't enough broadcast licenses to go around" and because high auction prices for DBS licenses show that they too are "scarce." Regulation should remedy the market's failure to achieve "program diversity" or produce "certain types of programs that are under-provided" by the standard of a properly functioning market (Hundt, 1996c). Opponents of regulation argue in return that competition obviates the need for regulation and provides all the diversity consumers want, mirroring the FCC's television deregulation order of 1984 (Quello, 1995). Recent FCC statements hail the competitive benefits that LMDS, a wireless alternative to cable television, will bring to the video market (FCC, 1997b). However, LMDS "[l]icensees are specifically on notice that the FCC may adopt public interest requirements…[possibly] including a 4% to 7% set aside of capacity for noncommercial educational and informational programming" (FCC, 1997b). Similarly, regulators welcome "the increasing competition among broadcast, cable, DBS, LMDS, wireless cable…[that] will be selling their wares to 100% of the country," but still find it "reasonable to put all media under some obligation to serve the public interest" through quantified programming obligations (Hundt, 1996a). Policies with very specific aims like political, educational, or children's programming are also often justified on the broad economic ground of market failure.

Given competitive developments in the video market and the emphases of new programming rules, however, it is unlikely that justifications based on program diversity and market failure are either persuasive or particularly relevant in debating the latest cycle of public interest regulation.

NEW VIDEO COMPETITION AND THE MARKET FAILURE JUSTIFICATION

Market failure generally refers to inefficiency created by externalities, public goods, or structural and informational impediments to efficient gains from trade. Broadcast regulators and policymakers have been most concerned with failures that prevent or deter licensees from providing programs that satisfy viewers' diverse preferences (Owen & Wildman, 1992). The existence of some market failures relative to the competitive optimum does not, in video or other markets, mean regulation is warranted. Whether it is depends in part on the magnitude of the market failure and on regulators' ability to identify both the failure and a promising remedy. Few markets function perfectly, but superior alternatives are often not feasible.

Sources of Video Market Failure

There are both informational and structural reasons that mutual, or at least net, gains from trade may be left unrealized in the video programming market. The fundamental informational difficulty is that video distributors may receive inaccurate signals about audience preferences. Structural factors relating to competition, costs, and channel capacity may limit the degree to which video distributors, whether pay-TV operators or conventional broadcasters of free programs, have individual incentives to respond efficiently to those preferences even if discernible.

Conventional broadcasters face a serious informational hurdle because viewers do not pay directly for broadcast television. When viewers freely consume as public goods programs advertisers sponsor, their preferences register only coarsely and indirectly: How many people watch a given program? The larger the audience, the more advertisers pay for time during a particular show and the more likely the broadcaster is to retain and increase such programming. In a properly functioning market, however, merely counting heads is a recipe for inefficiency. Holding costs constant, the market should produce the goods consumers value most, as communicated through the prices they are willing to pay. It maximizes that value—and increases consumer welfare—to produce 10 units of a good that consumers value at $10 per unit rather than to produce 50 units of a good that consumers value at $1 per unit. By counting heads, however, the conventional broadcast market will produce the program valued by 50 viewers at $1 instead of the program valued by 10 viewers at $10—the result of consumers' inability to communicate preferences through prices. Except where program popularity and preference intensity correspond, broadcasting will suffer this market failure.

Another possible failure in the broadcasting market is a consequence of programming incentives that may arise under limited competition. Early economic models showed how competition among the few VHF stations in a given market might produce less diversity of programming than would have been economically rational were the stations under unified, monopoly ownership. Steiner (1952) showed that where (a) programming preferences are sufficiently strong that viewers will not watch substitutes for their preferred formats, and (b) consumer preferences are skewed toward limited types of programs, oligopolistic competitors will rationally duplicate each other's programming to a greater extent than would occur if the channels were under monopoly control. Competing channels will duplicate the most popular kind of programming so long as their individual shares of the advertising revenues from such programming are higher than the advertising revenues to be gained by switching to the next most popular format. The result could be program homogeneity that is economically rational for each broadcaster but that leaves segments of the viewing market underserved (Brennan, 1983; Steiner, 1952).

If the channels were brought under joint ownership, however, then a more efficient program mix could potentially be achieved.[2] Assuming viewers of the most popular program type would all watch just one of the channels if the other two stopped showing what they wanted, the joint owner can place the most popular kind of programming on one channel and gain all the advertising revenue for that programming, but also have the other two channels available for the next most popular shows and their associated revenues.

Beebe (1977) demonstrated that even where Steiner's assumptions about preferences are relaxed and competitive duplication diminishes, a limited-capacity broadcast market functions inefficiently. If consumers do not limit their viewing to first-choice programming, a monopolist has incentive to seek common-denominator programming that most viewers are willing to watch rather than to

---

2. Joint ownership of course raises problems of its own in concentrated markets and was barred by the FCC's rule that broadcast licenses in a given market could not be jointly held.

produce a variety of programs that increase costs without expanding viewership. The monopolist might thereby maximize audience size yet fail to deliver many viewers their first-choice programming. Under competition, however, any individual channel needs to provide not only shows that the audience prefers to turning off the TV, but also programs viewers prefer to offerings by competing channels. If a channel offered only common-denominator programming it would lose viewers to competitors who offer programs that various audience segments prefer to the common-denominator type. Competition may thereby offer more viewers their first choices than does monopoly in some circumstances. The result in a limited-capacity market for free broadcasting may thus be that monopoly maximizes the number of viewers who get some satisfaction from broadcasting, whereas competition increases average satisfaction for any individual viewer. Which is preferable from a welfare standpoint is indeterminate in the abstract, but neither is likely to be optimal (Owen & Wildman, 1992).

Paid video services like cable and DBS suffer less than conventional broadcasting does from either of the preceding market failures, although the pay-TV market still exhibits biases that impede efficient programming decisions. Spence and Owen (1977) and Wildman and Owen (1985) demonstrated that a shift from a purely free broadcasting system to a market that contains pay-TV services is generally welfare enhancing. One drawback of free, advertiser-supported television is that some programs that are more valuable to viewers than to advertisers will not be produced because they are unprofitable. In some such cases a gain from trade is lost: Consumers would pay enough to make the show profitable, but there is no mechanism for such payments in the conventional broadcast system. Noll, Peck, and McGowan (1973) estimated that the lost gains are considerable because viewers are generally willing to pay more per program than advertisers are willing to pay for viewers. A pay-TV system lessens this inefficiency by partially accounting, through subscription prices, for the strengths of viewer preferences.

There are both positive and negative welfare effects from introducing consumer prices into the broadcasting market. Inefficiency results because the marginal cost of distributing pay-TV to an additional viewer is zero, yet that viewer will not receive the program unless he or she pays the (nonzero) price charged by the distributor. Conventional broadcast, in contrast, is priced efficiently at its marginal cost of zero. Some net welfare gains are thus foregone through pay-TV. The pricing inefficiency of pay broadcasting is common in markets for products with high fixed costs of production and would diminish if pay systems could discriminate in prices charged to different viewers.

On the positive side, pay-TV likely generates more diverse programming than free broadcasting does. The bias against minority tastes and in favor of large audiences endemic to advertiser-supported programming decreases where smaller audience groups can bid through prices for their desired formats. Cable and DBS operators control multiple channels and thus have more incentive and ability to diversify program offerings. Pay-TV will still fail to produce some programming for which the consumer benefits are greater than production costs because willingness to pay will sometimes be concentrated in relatively few viewers. But the situation is better than under an advertiser-supported regime where strong viewer preferences cannot be communicated at all through price. (Owen & Wildman, 1992).

The Effect of New Competition on Video-Market Failures

As noted, recent public interest rules continue to be justified by reference to market failure in video distribution. Strong competition among increasing numbers of video distributors should, however, be expected to limit the kinds of market failure with which public interest regulation has conventionally been concerned. Indeed, three factors significantly differentiate today's video market from the market of even 10 years ago: (a) the increased number of competing broadcasters, (b) an increase in the number of subscription video providers, and (c) increased channel capacity. Each factor reduces the magnitude of video market failures.

MORE COMPETING BROADCASTERS. As the number of channels grows, failures in the broadcast market diminish because individual competitors have stronger incentives to improve and differentiate their programming. Even in circumstances where monopolisitic competition produces excessive duplication, new entrants eventually find that further sharing the market for a popular program type is less profitable than having a larger share of the market for less popular programming. (Beebe, 1977; Steiner, 1952). Accordingly, as the number of competitors in the broadcast market increases, so too does the range of programming that becomes individually rational for competing channels to provide. Some inefficiency will remain due to the inability of broadcasters to account for preference intensities, but that market failure also diminishes because the range of preferences satisfied through the market increases. Another possible negative consequence of competition is that quality may suffer if it is correlated with a program's cost. As competitors divide audience share, the revenues for any given program may decline and drive channels to seek lower cost formats. On the other hand, quality will also be a dimension in which channels compete.

The number of competing broadcasters has in fact grown with technological advances in spectrum usage. The number of television stations nationwide expanded from 1,197 in 1985 to 1,532 in 1995 (Compaine, 1997), and Fox has succeeded in forming a fourth major network. Thus, even putting aside incumbent cable operators, DBS operators, and other possible entrants, the increased number of conventional broadcasters alone helps to overcome failures in the commercial broadcasting market.

MORE SUBSCRIPTION VIDEO DISTRIBUTORS. Of particular importance to the functioning of the current video market is the increasing number of competing subscription video program distributors (MVPDs) as DBS and other services are beginning to compete with cable systems. Most major markets now have access not only to service from an incumbent cable operator, but also to a choice of DBS services from competing providers like DirecTV, Echostar, and Primestar. The programming incentives of MVPDs differ from those of single channel operators and will likely lead to greater efforts to identify and produce programming for diverse preference groups in the market. A MVPD might try to stimulate market demand by adding variants of popular formats, perhaps by running similar formats on different channels at different times. But such schemes do not change a multichannel operator's incentive to satisfy demand for as many types of programming as possible and to avoid unnecessary duplication of formats.

When consumers can choose between different channel menus, there is stronger market incentive for operators to (a) choose the mix of programming that will attract the most subscribers to their individual systems, and (b) profitably differentiate their individual program mixes from those of competitors, thereby increasing diversity. The high fixed costs of programming suggest that product differentiation will be a particularly important dimension of video competition (Owen & Wildman, 1992). This is not to say the multichannel market is perfect by any means. Prices for bundled programming do not fully communicate preferences for individual programs within the bundle. Some pay-TV is partly advertiser-supported. Advertisers and viewers may have different preferences about what should be aired and that difference may affect which channels get carried by a given system. But economic analysis generally reveals that competition in pay-TV increases diversity and satisfies a broader range of viewing preferences (Spence & Owen, 1977; Wildman & Owen, 1985).

INCREASED CHANNEL CAPACITY. Economic studies also demonstrate that an increase in channel capacity increases consumer welfare (Owen & Wildman, 1992). As already discussed, the increase in individual broadcast channels expands the range of programming licensees will offer. Similarly, the more channels controlled by a multichannel operator, the greater the range of programming and program packages the operator can offer to consumers.

In the past decade, average channel capacity for cable systems has expanded substantially. More than half of all U.S. cable subscribers receive at least 54 channels and only about 3% of subscribers now receive fewer than 30 cable channels (Compaine, 1997). DBS systems offer 120 video channels and promise even more. Fifty or 100 channels give distributors enormous room rationally to include programming far down the ranking of collective consumer preferences and reduce the negative effects of divergent consumer and advertiser preferences. Communication of consumer preferences through the video market is also enhanced by the increasing number of program packages an MVPD can offer.

Evidence of Competitive Effects

Empirical evidence confirms that there is strong competition among video systems for viewers and advertising dollars. Although some potential competitors like telephone systems and "wireless cable" operators remain constrained by high costs or technological factors, competition for viewers has been strong in the video market since 1984. Cable systems grew through the 1980s and can now claim two thirds of U.S. households as subscribers (FCC, 1995). Videocassette recorders also became ubiquitous in the 1980s and consumers by 1989 were spending more on video rentals than businesses were spending for advertising on major broadcast networks (Setzer & Levy, 1991). In the face of such competition, broadcasters saw prime-time market share drop from 89% in 1980 to 64% in 1991. The major TV networks have seen their share of prime time viewers drop from 75% in 1985 to 49% in 1996 (Compaine, 1997).

Video competition has only grown in recent years. With the advent of commercially viable DBS systems, the number of satellite subscribers grew from fewer than 70,000 in 1993 to over 5 million by 1997 (Krattenmaker, 1997). The recent drop in dish prices coupled with the vast packages that can be offered over DBS will put increasing pressure on competitors. DBS operators estimate that at least one third of their subscribers were not formerly cable customers, indicating that DBS exerts substantial new competitive pressure on conventional broadcasters as well as on cable operators.

In addition, more programming has become available for competing operators to offer their customers. In 1980 there were only 28 commercial program producers for cable operators to choose from. The number of cable networks doubled to 56 by 1985, however, and by 1996 nearly tripled again to 162 (Compaine, 1997). Those networks compete vigorously for carriage and enable cable and DBS systems to offer subscribers a broader range of programming.

The number and type of competing video distributors thus fundamentally distinguish today's video market from that of just a few years ago. There are now stronger competitive pressures, more efficient incentives, and much greater ability for competitors to diversify and improve programming. The point is not that video market failures have ceased to exist; There remain intrinsic flaws. However, given the stronger incentive and ability for video competitors themselves to correct such failures, they are likely to be much less common or significant.

POLICY IMPLICATIONS

The recent market developments in video distribution have consequences for the public interest debate. First, the case for rules designed to enhance program diversity or remedy inefficient programming gaps is greatly weakened. Second, policy objectives that remain even as the video market works better for consumers likely have little to do with market failures and might more usefully be debated in terms of their noneconomic motivations.

Public Interest Regulation as a Remedy for Market Failures

Even in the relatively simple case of monopolistic competition among VHF stations, correcting for market failures is difficult. To begin with, regulators must distinguish programs the market does not want from programs not produced because of market failures. A strong signal from the public may reflect the view of many or the concerted view of an organized few. Are the "over 20,000 parents and other individuals that sent...letters and e-mails urging the FCC to take action" on children's television evidence of market failure, given that there are over 60 million television-watching households in the United States (Hundt, 1997)? Moreover, broadcasters might omit several types of desired programming because they duplicate popular formats or because revenues undercount consumer benefits. Which should receive regulatory attention? How should the presence of video or nonvideo substitutes for omitted programming be factored into the analysis? Difficult empirical questions about demand structure nettle any such inquiry. The FCC has itself found its programming guidelines to be ineffective while imposing significant administrative costs (FCC, 1984).

The Weakened Economic Case for Programming Rules in the Current Market

The fact that increasingly diverse preferences are satisfied through growing channel capacity and competition makes programming rules designed to increase diversity even less tenable than in the past. Because current video competition greatly expands the range of consumer demands that distributors have the incentive and ability to satisfy, public interest rules are unnecessary to achieve mere program diversity.

Rules that attempt instead to identify specific categories of programming that are inefficiently lacking also become more difficult to justify on an economic basis. In principle, the purpose of public interest regulation has been to affect the video market on the margin by replacing programs the broadcaster would have chosen with programming mandated by rule. The recent changes in the video market have significantly reduced the probability that a programming gap is due to market failure. The absence of a particular kind of programming may instead signal lack of demand, either because consumers simply do not want such programs or because demand is met by alternative products. Videocassettes, CD-

ROMs, the Internet, or good old-fashioned printed media might be winners in a given market segment.

The video market has in the past few years become capable of serving small market segments whose demand previously could not be discerned or served by video distributors. Any mandated categories might come at the expense of some such diverse programming. If regulation does not remedy a market failure, substituting required programming for that which the market produced will likely reduce current consumer welfare. How large a welfare loss will occur depends in part on availability of video and nonvideo substitutes for the removed programs. However, if one of the achievements of the current video market is to serve minority tastes that have few substitutes, programming rules risk being counterproductive from a consumer standpoint.

THE SCARCITY RATIONALE AND QUID-PRO-QUO PROPOSALS. Scarcity of useful frequencies does not supply an independent justification for programming rules (Krattenmaker & Powe, 1994; Thorne, Huber, & Kellogg 1995). Most resources are scarce; what matters is that they be allocated efficiently. The question then is whether there is a market mechanism that communicates the information necessary for producers and consumers to allocate scarce conduit efficiently among program options. The scarcity rationale thus boils down analytically to the market failure justification. As already discussed, current video competition diminishes market failure concerns, in part by reducing scarcity itself through technologies that expand channel capacities and usable frequencies.

The FCC often relies on a version of the scarcity rationale to characterize programming rules as a fair return to the public for some benefit conferred on a particular class of licensees. Former Chairman Hundt has described children's television obligations as just return for broadcasters: free digital spectrum (Hundt, 1997). FCC Commissioner Ness has suggested that because Congress gave broadcasters carriage rights and scarce channel positions on cable systems, "it is fair for [the FCC] to continue to establish and enforce public interest obligations such as a commitment to educational and informational programming" (Ness, 1996). Such statements are politically understandable given that most spectrum licensees now pay for their rights at auction whereas broadcasters continue to occupy their frequencies for free. There is intuitive appeal to the notion that broadcasters should also pay their dues one way or another.

Even if some broadcasters have received a windfall from free spectrum or carriage rights, however, that gain does not itself justify any particular policy. In some cases a regulatory windfall might justify imposing the costs of a good policy on broadcasters, but it has nothing to do with whether or not a given policy is wise to begin with. If a programming rule will reduce consumer welfare by replacing desired programming with some other type, the loss of satisfaction to viewers is not compensated by the fact that broadcasters are now somehow paying for their spectrum rights. Therefore, despite the pleasing, Robin Hood-like character of such quid pro quo proposals, they might well compound, rather than offset, the public costs of the policies for which they are being asked in exchange.

Implications for Specific Programming Rules

Market developments that have rendered much programming regulation unnecessary have certainly not cured all problems with video media. Serious issues may remain relating to program quality, the social effects of certain programming, or the use of video media to achieve social goals that have nothing to do with consumer desires. Current discussion of broadcast regulation often focuses less on diversity itself than on issues like the effects of televised violence or the potential benefits of increased political coverage or higher quality children's programming. It is perhaps no more correct to say that regulation's noneconomic objectives will be achieved by the market than it is to say that the video market's economic performance needs a boost from regulation.

Consider children's programming. The issue in the debate over the FCC's "kidvid" rules, implemented pursuant to the Children's Television Act (1990), is not the mere availability of programming aimed at children. Broadcasters have long devoted many weekly programming hours to children. Nor is the issue that children themselves were dissatisfied with what the market brought them. Rather, the debate is over the judgment that more educational content should be included in programming aimed at children. This debate is, at root, not principally an economic one.

Similarly, increased political coverage has been presented as a possible aid to campaign finance reform and has been discussed in terms of the benefits it could have for the quality of political discourse and participation in the United States (Sunstein, 1993). Whether such benefits warrant drafting broadcasters and other licensees into the regulatory effort has nothing to do with whether consumers currently demand such programming from video providers.

To be sure, there are economic angles to regulation targeted at political or children's programming, although they have little to do with the market failures with which public interest rules have conventionally been concerned. Moreover, the economic aspects should not distract the debate from the more fundamental noneconomic goals on which the validity of such regulation must ultimately be judged. For example, one could argue that kidvid rules constitute economic regulation because they address the market's failure to discount the irrational preferences of children in favor of the preferences of parents who cannot be constant gatekeepers to the television. Even if it is true that parental preferences diverge from those of children, this is not a conventional case of market failure. It is more a statement that even if the market operates perfectly and direct consumer preferences are fully communicated, those preferences are themselves somehow the "wrong" ones. This situation is not unique to children or to the video market. There are many cases in which one might want others to change their personal consumption decisions for some greater good. The questions of why children's preferences may be irrational and of whether parental preferences can and should supplant those of children are more fundamental to the policy debate than are issues of market failure.

There are, of course, sometimes externalities from "bad" or irrational preferences. Perhaps a steady diet of cartoons and toy-based programs creates uneducated, unproductive children. However, this possibility does not distinguish children's television from other programming that arguably has bad public effects but is not addressed by regulators. The distinguishing factor in the decision to regulate children's television must therefore be some broader rationale in which any public-economic concern is merely embedded—perhaps that the great power of mass media will foster particular social benefits by carrying better, more educational content to children. That determination, and associated judgments about what constitutes "quality" or "educational" programming, are independent from economic considerations of market failure and consumer welfare.

It has also been suggested that as viewers begin to watch higher quality children's programming or political coverage, they will discover they prefer it to other fare and be better off (Hundt, 1997). This does not mean, however, that rules requiring such programming are correcting flaws in the video market. It is probable in all markets that preferences evolve with consumption. The fact that current production and consumption decisions cannot anticipate such dynamic preferences is a cognitive limitation, not a market failure. Moreover, if current consumption affects future preferences, then rules affecting current output must be selected based on what policymakers believe those future preferences should be. Rather than pursuing the conventional goal of maximizing welfare, regulators are making the normative decision of what the welfare function will be. It may be right that society will be better off because children and parents watch more edifying television, but that, too, seems more a social judgment than an economic one and should be debated accordingly.

Public interest rules promoting political coverage or children's programming thus do not fit the conventional mold or justifications for broadcast regulation. It is probably inaccurate and irrelevant to present them as remedies for market failure. This does not mean the rules lack merit. It is similarly inapposite to oppose such regulation on grounds of increased market competition. Rather, their merit depends on the answers to difficult, noneconomic questions. Current public interest policies raise theoretical issues about competing social values and the proper role of government in selecting those values. They raise practical questions of how regulators will determine, for example, whether the content of a particular program is of a sufficiently high quality or educational nature to satisfy the FCC's children's programming rules. These various questions are certainly not ignored by scholars or policymakers. (Campbell, 1997; Krattenmaker & Powe, 1994).

However, they might become more central and less obscure in the public debate if regulators unbundled them from the conventional, and increasingly irrelevant, economic language of market failure and product diversity.

CONCLUSION

Were the FCC to revive the kinds of programming guidelines repealed in 1984, the surrounding debate would usefully focus on the market failures, consumer welfare, and program diversity concerns that have historically been central to public interest regulation. The beneficial developments in video competition and capacity counsel against a reversion to programming rules designed merely to enhance diversity and remedy remaining flaws in the market. For the most part, however, neither Congress nor the FCC has significantly reversed the deregulatory decisions of the 1980s. Public interest regulation today ignores many of the old categories of mandated programming in favor of a few specific targets that are motivated less by economic than by social or political goals.

The current policy debate, however, often discusses those targets in economic terms conventionally at the center of public interest regulation. Economic analysis of the current video market suggests that those terms of debate are not particularly relevant or useful for examining current public interest rules. It is unlikely that children's television rules, for example, remedy any actual failure of today's more competitive video market. If increasing conventional measures of consumer welfare were the only reason for those rules, the rules should be repealed. Those rules, however, either address different kinds of economic goals or, more fundamentally, noneconomic goals that are beyond the scope of even perfectly functioning markets. Market analysis will likely tell us little about the wisdom of such regulation. To allow clearer evaluation of current policy, therefore, the public interest debate would usefully shift further away from conventional economic terms and toward more direct discussion of underlying social and political objectives.

ACKNOWLEDGEMENTS

I wish to thank Angela Campbell, Ben Compaine, the Editors, and participants in the 1997 Telecommunications Policy Research Conference for helpful comments and suggestions.

REFERENCES

Ashbacker Radio Corp. V. FCC, 326 U. S. 327 (1945).
Bremer Broadcasting Co., 2 F.C.C. 79 (1935).

Brennan, T. (1983). Economic efficiency and broadcast content regulation. *Federal Communications Law Journal, 35,* 117–29.

Beebe, J. H. (1977). Institutional structure and program choices in television markets. *Quarterly Journal of Economics,* September *91,* 15–37.

Children's Television Act of 1990, codified at 47 U.S.C. 303.

Compaine, B. (1997). *Reassessing Video Competition: Has Technology or Regulation Made a Difference?* Paper presented at the 25th Annual Telecommunications Policy Research Conference, Washington, D.C.

Campbell, A. (1997). *Lessons from Oz: Quantitative guidelines for children's educational television.* Paper presented at the 25th Annual Telecommunications Policy Research Conference, Washington, D.C.

Citizens Committee to Save WEFM v. FCC, 506 F.2d 246 (D.C. Cir. 1974).

FCC v. Sanders Bros. Radio Station, 309 U. S. 470 (1939).

FCC v. WNCN Listener's Guild, 450 U. S. 582 (1981).

Federal Communications Commission. (1946). *Report, public service responsibility of broadcast licensees.* Washington, D.C.: Author.

Federal Communications Commission. (1960). *Programming policy statement.* 44 F.C.C. 2316.

Federal Communications Commission. (1965). *Public notice, policy statement on comparative broadcast hearings.* 1 F.C.C.2d 393.

Federal Communications Commission. (1976a). *Amendment to Section 0.281 of the Commission's rules: delegations of authority.* 59 F.C.C.2d 491.

Federal Communications Commission. (1976b). *Memorandum opinion and order, development of policy re: changes in the entertainment formats of broadcast stations.* 60 F.C.C. 2d 858.

Federal Communications Commission. (1982a). *Inqui.y into the development of regulatory policy in regard to DBS.* 90 F.C.C.2d 676.

Federal Communications Commission. (1982b). *Memorandum opinion and order, application of satellite television corporation for authority to construct an experimental direct broadcast satellite system.* 91 F.C.C.2d 953.

Federal Communications Commission. (1983). *Notice of proposed rulemaking, revision of programming and commercialization policies, ascertainment requirements and program log requirements for commercial television stations.* 94 FCC 2d 678.

Federal Communications Commission. (1984). *In the matter of the revision of programming and commercialization policies, ascertainment requirements, and program log requirements for commercial television stations.* 98 F.C.C. 2d 1076.

Federal Communications Commission. (1986). *Notice of proposed rulemaking, subscription video services.* Dkt. No. 85-305. 51 Fed. Reg. 1817.

Federal Communications Commission. (1987). *Report and order, subscription video.* 2 F.C.C. Rec. 1001.

Federal Communications Commission. (1995). *Second annual report, Annual assessment of the status of competition in the market for the delivery of video programming.* FCC No. 95-61.

Federal Communications Commission. (1997a). *In the matter of rulemaking...to establish rules and policies for LMDS and for fixed satellite services.* 1997 FCC LEXIS, FCC Dkt 97-82, March 13, 1997.

Federal Communications Commission. (1997b). *Statement regarding commission adoption of LMDS service and auction rules.* 1997 FCC LEXIS 1258, March 11, 1997.

Federal Radio Commission. (1928). *Second annual report.* Washington, D.C.: Author.

Federal Radio Commission. (1929). *Third annual report.* Washington, D.C.: Author.

Greater Boston Television Corp. v. FCC, 444 F.2d 841 (D.C. Cir. 1970).

Hundt, R. (1996). Speech, Broadcasting and Cable Interface Conference, Sept. 24, 1996. (1996 FCC LEXIS 5322).

Hundt, R. (1996a). Speech, Digital Convergence: Reshaping the Media, September 30, 1996. (1996 FCC LEXIS 5406).

Hundt, R. (1996c). Speech, American Bar Association, March 28, 1996 (1996 FCC LEXIS 1504)

Hundt, R. (1997). Speech, Kid's TV: The Impossible Dream has Become Inevitable, September 18, 1997. (1997 FCC LEXIS 5088)

KFKB Broadcasting Assn. v. Federal Radio Commission, 47 F.2d 670 (1931).

Krattenmaker, T. J. (1997). *Telecommunications law and policy* (2nd ed.). Durham, NC: Carolina Academic Press.

Krattenmaker, T. J., & Powe, L. A., Jr. (1994). *Regulating broadcast programming.* Cambridge, MA: The MIT Press and The AEI Press.

NBC v. United States, 319 U. S. 190 (1942).

Ness, S. (1996), Speech, New York Chapter, Federal Communications Bar Association, June 4, 1996. (1996 FCC LEXIS 2954)

Noll, R. G., Peck, M. J., & McGowan, J. J. (1973). *Economic aspects of television regulation.* Washington, DC: Brookings Institution.

Office of Communication of the United Church of Christ v. FCC, 359 F.2d 994 (D.C. Cir. 1966).

Owen, B., & Wildman, S. (1992). *Video economics.* Cambridge, MA: Harvard University Press.

Palmetto B'casting Co., 33 F.C.C. 250 (1962).

Quello, J. H. (1995). Speech, National Association of Broadcasters Children's Television Symposium, September 21, 1995. (1995 LEXIS 6256)

Scroggin & Co. Bank, 1 F.C.C. 194 (1935).

Setzer, F., & Levy, J. (1991). Broadcast Television in a Multichannel Marketplace, (FCC OPP Working Paper No. 26). (1991 FCC Lexis 5170)

Spence, M., & Owen, B. (1977). Television Programming, Monopolistic Competition, and Welfare. *Quarterly Journal of Economics, 91*, 103–26.

Steiner, P. O. (1952). Program patterns and preferences and the workability of competition in radio broadcasting. *Quarterly Journal of Economics, 66,* 194–223.

Sunstein, C. R. (1993). *Democracy and the Problem of Free Speech.* New York: The Free Press.

Teeter, D. L., Jr., & Le Duc, D. R. (1992), *Law of mass communications* (7th ed.).

Thorne, J., Huber, P. & Kellogg, M. (1995). *Federal broadband law.* Boston: Little, Brown.

Wildman, S. S., & Owen, B. (1985). Program competition, diversity, and multichannel bundling in the new video industry. In E. M. Noam (Ed.), *Video media competition: Regulation, economics and technology.* New York: Columbia University Press.

Lessons from Oz:
Quantitative Guidelines for Children's
Educational Television

Angela J. Campbell
Georgetown University Law Center

In the summer of 1996, more than 20 years after first declaring that television stations had an obligation to provide educational programming for children, the Federal Communications Commission (FCC) finally answered the question "How much?" In short, the FCC adopted a guideline of 3 hours per week. Although 3 hours a week may not seem like much, it took no less than a White House summit in an election year to obtain this result. In addition to establishing the guideline of 3 hours, the FCC defined what programming would qualify and established a number of procedures to increase public monitoring. Broadcasters agreed not to challenge the constitutionality of the FCC's decision.[1]

To anticipate the success of the FCC's newly adopted quantitative approach, this chapter examines the experience of Australia, which has long required commercial television stations to air a specific amount of children's programming. It begins with an overview of the reasons why it is useful for the United States to look to the Australian experience, as well as some of the limitations. It then sets out the history

---

1. *See* Supplemental Comments of the National Association of Broadcasters, MM Dkt. 93-48 (filed July 29, 1996) at 2. This chapter does not address the constitutionality of the FCC's order. For a discussion of this issue, compare Reed Hundt and Karen Kornbluh, *Renewing the Deal Between Broadcasters and the Public: Requiring Clear Rules for Children's Educational Television*, 9 HARV. J. L. & TECH. 11, 20–22 (1996) with James J. Popham, *Passion, Politics and the Public Interest: The Perilous Path to a Quantitative Standard in the Regulation of Children's Television Programming*, 5 COMM. L. CONSPECTUS 1, 25–27 (1997).

of children's television regulation in both countries. The final section suggests lessons that the United States might draw from the Australian experience.

THE RELEVANCE OF THE AUSTRALIAN EXPERIENCE

Australia has established itself as a leader in children's television and as a result, a number of commentators have suggested that it provides a model for the United States.[2] Other factors also make Australia a useful case study for comparison. Australia and the United States both began as British colonies, and thus share the English language and a legal tradition based on the British common law.[3] The constitutions of both countries share common fundamental features.[4] One significant difference, however, is that Australia's Constitution contains no explicit guarantee of freedom of speech comparable to the First Amendment to the U.S. Constitution.[5]

Australia and the United States have similar structures and processes for regulating broadcasting. In both countries, television stations are licensed for a limited, but renewable license term, and licensing is administered by a federal administrative agency—the FCC in the United States and the Australian Broadcasting Au-

---

2. *See, e.g.*, EDWARD L. PALMER, TELEVISION & AMERICA'S CHILDREN: A CRISIS OF NEGLECT 147–48 (1988); CASS R. SUNSTEIN, DEMOCRACY AND THE PROBLEM OF FREE SPEECH 85 (1993). This chapter is premised on the assumption that the Australian standards have worked as intended. Australian stations have complied with the quantitative requirements. See Australian Content and Children's Television Compliance Figures, ABA UPDATE, Nov. 1996, at 5–9 (finding that with one exception, the networks exceeded the minimum requirements for 1991 through 1995). Moreover, the ABA believes that the regulations have met the objective of "providing [children with] access to a variety of quality television programs made specifically for them." *See, e.g.*, *A Brief History of Standard*, in ABA UPDATE, March 1995, at 11. Finally, Australian children seem to have access to a wider variety of programming designed for them than U.S. children have had in recent years. *See Children's Television Standards*, ABA UPDATE, Feb. 1997, at 16 (describing critically acclaimed Australian children's programs and listing programs classified by ABA in 1996).

3. *See* MARY ANN GLENDON ET AL., COMPARATIVE LEGAL TRADITIONS 279–280 (1985).

4. William Rich, *Converging Constitutions: A Comparative Analysis of Constitutional Law in the United States and Australia*, 21 FEDERAL LAW REVIEW 202, 203 (1993).

5. The Australian Constitution, enacted in 1901, is said to blend elements of the British system of government with features derived from the U.S. Constitution. The founders of Australia, however, specifically rejected the need for a Bill of Rights comparable to the U.S. Bill of Rights. Australia recognizes freedom of speech as an important aspect of its democracy, but generally relies on Parliament rather than the courts to protect that freedom. Neil Douglas, *Freedom of Expression Under the Australian Constitution*, 16 U. NEW S. WALES L. J. 315, 318–321 (1993).

thority (ABA) in Australia. License renewal in both countries is conditioned on some notion of serving the public interest.[6]

Moreover, both the United States and Australia have a long tradition of commercial broadcasting.[7] In both countries, commercial broadcasting is supplemented by a publicly supported network—the Public Broadcasting System (PBS) in the United States and the Australian Broadcasting Corporation (ABC) in Australia. However, households in the United States generally have more alternative sources of programming, such as cable television and Digital Broadcast Satellite (DBS), available to them.[8]

CHILDREN'S TELEVISION REGULATION IN AUSTRALIA.

History

Regulation of children's television in Australia dates to the introduction of television. The 1953 Royal Commission on Television, charged with making recommendations concerning the introduction of television into Australia, recommended the establishment of a Children's Advisory Committee to advise the licensing agency, then known as the Australian Broadcasting Control Board. The Control Board established a Children's Advisory Committee in 1956.[9]

---

6. Under Section 309(a) of the Communications Act, the FCC may grant an application for renewal of license only where it finds that the "public interest, convenience and necessity would be served" 47 U.S.C. § 309(a). The meaning of the public interest standard is spelled out in a variety of FCC reports and decisions and has changed over time. *See, e.g.*, Television Deregulation, 98 FCC 2d 1076, 1093 (1984) (renewal standard consists of obligation to address community issues with responsive programming and compliance with all other legal requirements). License renewal is contingent on the provision of "adequate and comprehensive service." MARK ARMSTRONG ET AL., MEDIA LAW IN AUSTRALIA at 155 (3d ed. 1995).

7. Australia has three commercial networks—seven, nine, and ten—that control nearly all stations and have stations in most markets. *See* ARMSTRONG at 156. Thus, three networks dominate commercial broadcasting in much the same way that the NBC, CBS and ABC networks traditionally dominated in the United States. In recent years, however, the number of stations in major U.S. markets has increased, permitting the formation of additional networks.

8. Cable television and DBS are available to the majority of U.S. television households. Approximately 62.1 million people subscribe to cable, 3.82 million subscribe to DBS, and less than 1 million subscribe to wireless cable systems. Annual Assessment of the Status of Competition in the Market of the Delivery of Video Programming, *Third Annual Report*, 12 FCC Rcd 4358, 4368, 4377, 4387-88 (1997). In Australia, pay television, which includes cable, MDS, and DBS, is just getting off the ground. There, five main pay-TV operators together have achieved only about 8.5% penetration. AUSTRALIAN PAY TV NEWS, Mar. 28–Apr. 11, 1997, at 11.

In 1968, the Control Board introduced an incentive within the Australian content standards for airing Australian children's programs. This effort proved unsuccessful in increasing the amount and quality of children's programs.[10] Thus, in 1971, the Control Board introduced a quota for Australian-made children's programs of 4 hours in 28 days.[11]

In 1977, the Control Board was replaced by the Australian Broadcasting Tribunal. The Tribunal was charged with determining "whether broadcasters should be allowed to regulate themselves in certain areas and, if so, what the minimum standards should be."[12] The Tribunal determined that self-regulation would not work in all areas of broadcasting because of the conflict between the profit incentive and social responsibility to children. Thus, the Tribunal implemented three new provisions.

First, the Tribunal established the C classification to indicate which programs are specifically designed for children. Because most programming shown on commercial television in Australia is classified according to the appropriate audience, the effect of this decision was to add a new classification.[13] The idea was

---

9. DISCUSSION PAPER—REGULATION OF CHILDREN'S PROGRAMS, *reprinted* in Australian Broadcasting Tribunal, KIDZ TV: AN INQUIRY INTO CHILDREN'S AND PRESCHOOL CHILDREN'S TELEVISION STANDARDS (1991) [hereinafter "KIDZ TV"] at 331, 334, 362.

10. *Id.* at 334. The Australian content standards have been revised several times, most recently in September 1995. ABA, AUSTRALIAN CONTENT, REVIEW OF THE PROGRAM STANDARD FOR COMMERCIAL TELEVISION, FINAL REPORT, Sept. 1995. The object of the standard is to "reflect[] a sense of Australian identity, character and cultural diversity by supporting the community's continued access to programs produced under Australian creative control." *Id.* at 1. Under the new standards, 50% of all programming must be made under Australian creative control and minimum amounts of first-release drama, children's programs (including children's drama) and documentary programs must be aired. *Id.* at 25–28.

11. KIDZ TV at 334–36. Over the next several years, the quota was increased to 10 hours.

12. AUSTRALIAN BROADCASTING TRIBUNAL, SELF-REGULATION FOR BROADCASTERS? A REPORT ON THE PUBLIC INQUIRY INTO THE CONCEPT OF SELF-REGULATION FOR AUSTRALIAN BROADCASTERS 7 (1977) [hereinafter "1977 REPORT"].

13. The present classification scheme has four categories: (a) General (G) programming cannot contain "any matter likely to be unsuitable for children to watch without the supervision of a parent," (b) Parental Guidance Recommended (PG) may contain adult themes but must remain suitable for children to watch under the guidance of a parent; (c) Mature (M) is recommended for viewing only by persons aged 15 or over; and (d) Mature Adult (MA) is suitable for viewing only by persons aged 15 or over. Factors considered in classification include violence, sex and nudity, language, drugs, and suicide. The Code of Practice delineates the time periods within which each classification may be aired e.g., MA programming may only be aired after 9:00 p.m. and before 5:00 a.m. News, current affairs, and live sporting programs are not classified. Commercial Television Industry Code of Practice, August 1993, § 2.

to provide parents and children with the information that certain programs have been designed with the interests and needs of children in mind. The current "G" classification merely indicates that a program is not considered unsuitable for children, but gives no indication of the actual nature of the program.[14]

Second, the Tribunal established the C time period from 4:00 p.m. to 5:00 p.m. Monday through Friday. The effect of permitting only C programs to be aired during C time was to require every station to air 1 hour each weekday of C-classified programming, for a total of 5 hours per week. Finally, the Tribunal established the Children's Program Committee (CPC) to formulate guidelines and criteria for C classification programs and to view all programs proposed for transmission in C time to determine whether they qualified. The CPC's seven members were appointed by the Tribunal, with four from the public and three from industry.

In 1982, the CPC recommended that the Tribunal adopt a series of standards. The Children's Television Standards (CTS) adopted in 1984 continued the requirement that stations broadcast only C programs from 4:00 p.m. to 5:00 p.m. each weekday. They also required that licensees show a minimum of 30 minutes of preschool (P) programs between 9:00 a.m. and 4:00 p.m. on each weekday.[15] Thus, the 1984 standards required a minimum of 7.5 hours of weekday programming specifically designed for school age and preschool children. The Tribunal promised to evaluate the standards after 2 years.

In February 1987, the Tribunal initiated an inquiry into the effectiveness of the CTS that led to revised standards effective January 1, 1990. The objective of the 1990 CTS is that "[c]hildren should have access to a variety of quality television programs made especially for them, including Australian drama and non-drama."[16] In the Report accompanying the revised standards, the Tribunal found that "[c]hildren have particular needs and interests in relation to television, however they do not have the purchasing power to make them attractive as a discrete group to advertising buyers." Thus, the Tribunal concluded that "industry initiative and market forces cannot be relied on, to the same extent as with the adult audience, to provide programs to meet the special needs of the child viewer."[17]

The 1990 CTS were intended to give broadcasters greater flexibility than the 1984 CTS in meeting the needs of children. Although the total amount of children's programming was not changed from the 1984 CTS, broadcasters were giv-

14. 1977 REPORT at 64.

15. Children's Television Standards (CTS) 3, *reprinted* in KIDZ TV at 313.

16. CHILDREN'S TELEVISION STANDARDS, AN INQUIRY CONDUCTED BY THE AUSTRALIAN BROADCASTING TRIBUNAL TO REVIEW THE CHILDREN'S AND PRESCHOOL CHILDREN'S TELEVISION STANDARDS, IP/87/4 (Nov. 1989), *reprinted* in KIDZ TV at 41 [hereinafter "1989 STANDARDS"].

17. DECISION AND REASONS (Nov. 1989), *reprinted* in KIDZ TV at 25, 27.

en more leeway in terms of scheduling. Instead of having to air only C programming between 4:00 p.m. and 5:00 p.m. weekdays, stations were permitted to establish their own C time within the C time bands, which included both a wider time period on weekdays and times on weekends and holidays.

The Current Standards

The 1990 CTS remain in effect today, despite the fact that major deregulation of broadcasting has occurred in Australia. The Broadcasting Services Act of 1992 replaced the Tribunal with the smaller Australian Broadcasting Authority (ABA) and dramatically reduced the amount of regulation in most areas. However, the 1992 Act specifically directed the ABA to determine standards for commercial television broadcasters with regard to programs for children and Australian content of programs.[18] The accompanying Broadcasting Services (Transitional Provisions and Consequential Amendments) Act 1992 carried over the existing standards for administration by the ABA.[19] Thus, for the present, the ABA continues to enforce the 1990 CTS with variations necessitated by changes to the Australian Content Standard adopted in 1995.

CTS 3 currently requires that each commercial television station air a minimum of 390 hours of children's programs per year. Each station must air at least 130 hours for preschool children (P programs) and 260 hours for school-age children (C programs). Each station is required to air P and C programs during specified time bands. Moreover, both P and C programs must be broadcast for a continuous period of time not less than 30 minutes. Each licensee must furnish to the ABA a schedule showing when it will broadcast P and C programs, and only programs broadcast according to the schedule will count toward the minimum. At least half of the C programs must be first-release Australian programs, and some component of these must fit the definition of Children's Australian Drama (CAD).[20]

CTS 2 defines a "C" program as one which:

1. Is made specifically for children or groups of children within preschool or primary school age range.
2. Is entertaining.

---

18. Broadcasting Services Act, 1992, § 122(2)(a).

19. Broadcasting Services (Transitional Provisions and Consequential Amendments) Act, § 21 (2) (1992).

20. CAD programming is programming classified as a C program that meets the Australian Content Standard's requirements for an Australian program, and is found by the ABA to "be a fully scripted screenplay or teleplay in which the dramatic elements of character, theme and plot are introduced and developed so as to form a narrative structure." CTS 11. Stations are required to air 24 hours of CAD programming in 1996, 28 hours in 1997, and 32 hours in 1998.

3. Is well produced using sufficient resources to ensure a high standard of script, cast, direction, editing, shooting sound, and other production elements.
4. Enhances a child's understanding and experience.
5. Is appropriate for australian children.

When these criteria were adopted, the Tribunal explained the basis for these criteria.[21] The first criterion—that programming be made specifically for children—was intended to alleviate "confusion between programs that children like and programs that are made specifically for them. They are not mutually exclusive but just because children enjoy certain types of programs it does not follow that they are specifically for them."

As to the second criterion that a program be "entertaining," the Tribunal explained, "a children's program can be drama or non drama, designed to educate or to be just good fun, but the aim of all C and P programs should be to entertain children." Regarding the third criterion that the program be well-produced, the Tribunal recognized that "although money does not necessarily equate with 'quality,' it is accepted that without the commitment of sufficient resources the high production values required would be difficult to achieve."

To meet the fourth criterion—enhancing a child's understanding and experience—program producers are expected to "understand the emotional, intellectual, social and other characteristics relevant to specific age groups of children and create and broadcast programs that address the specific needs and interests of those children." Finally, as to the last criterion, the Tribunal expressed concern that programs produced in other countries may assume knowledge and understanding of the country of origin that Australian children may not have and reminded producers to take into account the "multicultural nature of Australia."

Rather than leave it to the broadcaster to determine whether a program meets the CTS 2 criteria, the ABA makes this determination in advance of a program's airing. Refusal to classify a program as C or P does not mean that the program cannot be aired, but only that a licensee cannot count it toward the C or P quota. It can still air the program as a G program. The ABA monitors stations' compliance with the quota on an annual basis. Failure to comply with program standards is a breach of condition of license that could result in loss of the license.

CHILDREN'S TELEVISION REGULATION IN THE UNITED STATES

History

From the earliest days of broadcasting, service to all substantial groups in a community was considered an essential ingredient of a broadcast licensee's pub-

---

21. The following quotations are from the DECISION AND REASONS, in KIDZ TV at 25 and EXPLANATION OF STANDARDS, in KIDZ TV at 69.

lic interest responsibilities.[22] During the 1950s, when television first became widely available, stations aired a relatively large quantity and wide variety of children's programs.

In 1960, the FCC first recognized children's programming as a distinct program category.[23] During the late 1960s, however, stations began cutting back on the amount of children's programming, particularly on weekdays. In 1970, Action for Children's Television (ACT) brought attention to the dearth of children's weekday programming by filing a Petition for Rulemaking with the FCC. ACT requested, among other things, that broadcasters be required to provide daily children's programming totaling at least 14 hours per week.

The 1974 Policy Statement put broadcasters on notice that they have a "special obligation to serve children."[24] It explained that "because of their immaturity and their special needs, children require programming specifically designed for them." It concluded that "the use of television to further the educational and cultural development of America's children bears a direct relationship to the licensee's obligation under the Communications Act to operate in the 'public interest.'"

Noting that some stations provided no programming for children, it stated that it expected stations "to make a meaningful effort in this area." The FCC further found that "over the years, there have been considerable fluctuations in the amount of educational and informational programming carried by broadcasters—and that the level has sometimes been so low as to demonstrate a lack of serious commitment to the responsibilities which stations have in this area." Thus, in the future, the FCC warned, "license renewal applications should reflect a reasonable amount of programming which is designed to educate and inform—and not simply to entertain." The FCC added that licensees should serve both preschool and school-aged children and that they should schedule programming throughout the week. Although the FCC concluded that it was not necessary to adopt rules prescribing a set number of hours of children's programming, it warned broadcasters to take immediate action to meet their public service responsibilities and that it would evaluate their efforts in the future.

In 1978, the FCC issued a second Notice of Inquiry[25] and established a Task Force to assess compliance with the Policy Statement. The Task Force concluded

---

22. See, for example, Great Lakes Broadcasting Co., 3 F.R.C. ANN. REP. 32, 34 (1929), *rev'd on other grounds*, 37 F.2d 993 (D.C. Cir.), *cert. dismissed*, 281 U.S. 705 (1930).

23. The 1960 En Banc Programming Statement, which set out licensees' public interest responsibilities, included "programs for children" among the types of programming that a broadcast station would normally be expected to provide. *Report and Statement of Policy Re: Commission En Banc Programming Inquiry*, 44 FCC 2303, 2314 (1960).

24. Children's Television Report and Policy Statement, 50 FCC 2d 1, 5, *recon. denied*, 55 FCC 2d 691, *aff'd sub nom.* Action for Children's Television v. FCC, 564 F.2d 458 (D.C. Cir. 1977).

25. 68 FCC 2d 1344 (1978).

that broadcasters had failed to comply with the Policy Statement's programming guidelines.[26] Thus, it recommended that the FCC adopt mandatory program rules because of what it termed "market failure."[27] In response, the FCC proposed a range of policy options, including a requirement that each station air a minimum quantity of children's educational programming as a condition of renewal.[28]

The FCC failed to act on these proposals, however, until 1984. Over the strong dissent of Commissioner Rivera, the FCC rejected the imposition of any quantitative program requirements and weakened the already weak 1974 Policy Statement.[29] It concluded that a quantitative children's program requirement was unnecessary and would raise legal and practical problems.

Given the lack of responsiveness at the FCC, advocates for children's television turned to Congress.[30] Timothy Wirth, chairman of the House Subcommittee on Telecommunications from 1981 to 1986 introduced several bills that would have imposed specific quantitative requirements as a condition of license renewal.[31] These bills faced strong objections from the broadcasters.

After Wirth's election to the Senate, the new Chairman of the House Subcommittee, Ed Markey, was able to craft a compromise that ultimately led to passage of the Children's Television Act of 1988.[32] However, President Reagan pocket vetoed the measure.

In the next Congress, bills were introduced that ultimately led to the Children's Television Act of 1990 (CTA).[33] The CTA requires the FCC, in reviewing applications for license renewal, "to consider the extent to which the licensee

---

26. TELEVISION PROGRAMMING FOR CHILDREN: A REPORT OF THE CHILDREN'S TELEVISION TASK FORCE 2 (October 1979).

27. *Id.* at 75–80. Specifically, the staff recommended requiring 5 hours per week of educational or instructional programming for preschool children and 2.5 hours per week for school-age children that would be aired between the hours of 8:00 a.m. and 8:00 p.m. weekdays. *Id.* at 76.

28. 75 FCC 2d 138, 147–152 (1979).

29. Children's Television Programming and Advertising Practices, 96 FCC 2d 634 (1984), *aff'd sub nom.* Action for Children's Television v. FCC, 756 F.2d 899 (D.C. Cir. 1985).

30. For a fascinating account of the machinations leading up to the passage of the Children's Television Act of 1990, see Dale Kunkel, *Crafting Media Policy*, 35 AMERICAN BEHAVIORAL SCIENTIST 181 (1991).

31. *See, e.g.*, Children's Television Education Act of 1983, H.R. 4097, 98th Cong., 1st Sess., 129 CONG REC. H8080 (1983); Children's Television Education Act of 1985, H.R. 3216, 99th Cong., 1st Sess., 131 CONG. REC. E3763 (1985).

32. H.R. 3966, 100th Cong., 2d Sess. (1988); S. 2071, 100th Cong., 2d Sess. (1988). This bill would have required that each station serve "the educational and informational needs of children in its overall programming."

33. Pub. L. 101-437, 104 Stat. 996-1000 (1990) (codified at 47 USC §§ 303(a), 303(b), and 394).

has served the educational and informational needs of children through the licensee's overall programming, including some programming specifically designed to serve such needs."[34]

To implement the CTA, the FCC initiated a rulemaking proceeding.[35] A number of commenters asked the FCC to adopt quantitative processing guidelines along the lines the FCC had proposed in 1979.[36] However, the FCC rejected this suggestion, citing legislative history suggesting that Congress meant for no minimum criterion to be imposed.[37] The FCC also noted that the amount of children's educational programming necessary to comply would vary depending on a variety of circumstances. Thus, the FCC merely required stations to keep publicly available records describing their children's programming and to submit a summary of their program records with their applications for license renewal.

In the fall of 1992, when the Children's Television Act had been in effect for 1 year, Georgetown University Law Center's Institute for Public Representation and the Center for Media Education conducted a study of how stations were serving the educational and informational needs of children. They found that overall, television broadcasters were not making a serious effort to serve children's educational and informational needs.[38]

This study garnered attention in Congress and the press, prompting the FCC to launch an inquiry to determine whether it should change its enforcement of the CTA.[39] After consideration of the comments, the FCC issued a Notice of Proposed Rulemaking (NPRM) in May 1995.[40] The NPRM set forth three options for comment: (a) further monitoring; (b) a processing guideline of 3 hours per week, and (c) a program standard of 3 hours per week.

It took almost another year for the FCC to finally adopt rules. With the departure of one of the five commissioners in the Spring of 1996, the FCC was deadlocked two to two. An unusual public dispute developed between Chairman Hundt and Commissioner Quello.[41] Finally, after President Clinton convened a White House Summit, the broadcast industry backed off, and a compromise was reached. As

---

34. 47 USC § 303(a)(2).

35. Policies and Rules Concerning Children's Television Programming, *Notice of Proposed Rulemaking*, 5 FCC Rcd 7199 (1990).

36. See, for example, Comments of National Association for Better Broadcasting et al., MM Docket No. 90-570 (filed Jan. 30, 1991).

37. Policies and Rules Concerning Children's Television Programming, *Report and Order*, 6 FCC Rcd 2111, 2115 (1991), *recon. denied*, 6 FCC Rcd 5093 (1991).

38. A Report on Station Compliance with the Children's Television Act 3 (September 29, 1992)("CME/IPR Report").

39. Policies and Rules Concerning Children's Television Programming, *Notice of Inquiry*, 8 FCC Rcd 1841 (1993).

40. Polices and Rules Concerning Children's Television Programming, *Notice of Proposed Rulemaking*, 10 FCC Rcd 6308 (1995).

explained by the General Counsel of the trade association of independent broadcast stations:

> [B]roadcast executives invited to the White House hardly relished the prospect of being chastised by a popular president in such a public setting. Broadcast industry lobbyists also saw significant risk in further delaying what already appeared inevitable. If, as anticipated President Clinton was re-elected in November, he would appoint two new Commissioners, both of which likely would share his position on children's television. If anything, a FCC without former broadcaster Jim Quello might be inclined to adopt even more stringent requirements. Therefore, neither the broadcast industry nor the White House remained willing to leave the issue of children's television to the bickering, dead-locked Commission any longer.[42]

The FCC issued its new rules in August 1996, and no party appealed or sought reconsideration.

The Current Standards

Under the processing guideline adopted by the FCC, licensees who fall into Category A or B can have their licenses renewed by the FCC staff. Stations that have aired 3 hours per week of "core" programming, averaged over a 6 month period fall into Category A. Category B stations are those that have aired "somewhat less" than 3 hours of core programming that "have aired a package of different types of educational and information programming that demonstrates a level of commitment to educating and informing children that is at least equivalent to" 3 hours of core programming. Applicants that do not fit within either category will be referred to the FCC for a full opportunity to demonstrate compliance with the CTA.[43]

"Core" programming must meet six criteria:

---

41. See Popham, *supra* note 1, at 11–15; Edmund L. Andrews, *A Bitter Feud Fouls Lines at the FCC,* NEW YORK TIMES, Nov. 20, 1995, at D1; Chris McConnell, *Fight heats up over kids TV,* BROADCASTING & CABLE, July 22, 1996, at 12.

42. Popham, *supra* note 1 at 15. See also *FCC urged to accept White House Kidvid Compromise,* COMMUNICATIONS DAILY, July 30, 1996; Chris McConnell, *Kids TV accord reached,* BROADCASTING & CABLE, Aug. 5, 1996, at 5.

43. Policies and Rules Concerning Children's Television Programming, *Report and Order,* 11 FCC Rcd 10660, 10718-24 (1996). The General Counsel to the trade association of independent stations asserted that "[n]o station licensee is likely to test the 'Category B' option, much less opt for full Commission review." Popham, *supra* note 1, at 21. Thus, the practical effect of the guideline is that stations will offer no more and no less than 3 hours of core programming per week.

1. Have serving the educational and informational needs of children ages 16 and under as a significant purpose.
2. Specify the educational objective of the program and target audience in writing in the station's children's program report.
3. Be aired between 7:00 a.m. And 10:00 p.m.
4. Be regularly scheduled.
5. Be of substantial length (i.e., at least 30 minutes).
6. Be identified as educational children's programming when aired and such identification must be provided to program guides.

The first element involves a subjective judgment first, as to whether the programming is educational and second, whether educating is a "significant purpose" of the program. Unfortunately, the FCC rules give little guidance as to either inquiry. The rules adopted in 1991 had defined educational and informational programming as "any television programming which furthers *the positive development* of children 16 years of age and under in any respect, including the child's intellectual/cognitive or social/emotional needs." After concluding that this definition "does not provide licensees with sufficient guidance," the FCC modified the definition by replacing the phrase "positive development" with "educational and informational needs."[44] In other words, the FCC now defines educational and informational programming as programming that serves the educational and informational needs of children.

The FCC explains that the "significant purpose" criterion is designed to encourage programming that educates and informs but also is entertaining and attractive to children. However, it never explains how it will determine whether education is a significant purpose.

The FCC emphasizes that whether programming qualifies as educational does not depend on the topic or viewpoint. To determine whether a program meets the first criterion, the FCC

> will ordinarily rely on the good faith judgment of broadcasters, who will be subject to increased community scrutiny as a result of the public information initiatives. We consequently will rely primarily on such public participation to ensure compliance with the significant purpose prong of the definition of core programming, with Commission review taking place only as a last resort.

The second element—specifying the educational objective in writing—is intended to help licensees focus on children's specific educational needs as well as to help parents better understand and evaluate licensees' responses. Reporting the target audience is intended to make it possible to determine whether children of different ages are receiving adequate service.

---

44. *Order*, 11 FCC Rcd at 10696-98.

The third element—scheduling—is designed to ensure that educational children's programming is aired when children can watch it. The last three elements are designed to help parents and children find educational programming. Regularly scheduled, standard-length programming is more likely to be listed in program guides. The FCC reasons:

> If parents have the opportunity to know in advance that a particular program has an educational and informational focus, and when such programs will be shown, they can encourage their children to watch such programming and thereby increase audience, ratings, and the incentive of broadcasters to air, and programmers to supply, more of such programming.

Identification of programming as educational is intended to improve the public's ability to monitor licensees' efforts and to complain when broadcasters fail to meet their responsibilities.[45]

LESSONS FROM THE AUSTRALIAN EXPERIENCE

By examining the experience in Australia, we may be able to predict the difficulties likely to be encountered by the new FCC rules. Australia's experience suggests that a quantitative requirement can work, even where the amount is more than twice that in the FCC guidelines. Yet, some Australian programs have been refused classification as children's programs.

What Programming Counts

In determining what program qualifies as a C or P program, the Australians have found that failure to receive classification typically results from problems in three main areas: the program is not made specifically for children, does not meet the requirements of a quality production, or does not enhance children's understanding and experience.

The issue of age specificity, the ABA notes:

> has remained one of the most problematic in children's programming. The ABA is still being asked to assess programs which betray a lack of understanding of the child's point of view or which are written from an adult's

---

45. The FCC adopted several other provisions designed to increase public participation such as requiring licensees to designate a children's television liaison, to place reports on their children's programming in the station's public file on a quarterly basis, and to utilize a standardized reporting form. It also encourages licensees to file electronic reports that could be made available on the FCC's home page on the World Wide Web. *Id.* at 10690-95.

point of view...[T]here is a noticeable tendency for drama to be written for the top end of the age range or to deal with themes more suited to adolescents. Age specificity is the most difficult criterion to meet and is the reason most often given for rejecting a program for C classification.[46]

Thus, it cautions that programs produced for a family, preschool, or adolescent audience cannot be classified C.

The tendency to program for a broader audience is not surprising in light of the economics of television programming. Because commercial broadcasters make money by selling audiences to advertisers, they have an incentive to air programming that will attract the largest number of viewers desired by advertisers. Children are inherently less attractive to advertisers than adults because they make up a smaller part of the audience. Furthermore, children have less money than adults to spend on advertised products and are interested in fewer advertised products. Educational programming is further disadvantaged in that it is relatively expensive to produce and tends to be targeted to a narrow age range. The FCC found that:

the combination of these market forces...can create economic disincentives...Broadcasters who desire to provide substantial children's educational programming may face economic pressure not to do so because airing a substantial amount of educational programming may place that broadcaster at a competitive disadvantage compared to those who do very little.[47]

Although use of a guideline will likely ameliorate the problem of penalizing broadcasters who do present educational programming, it does not eliminate the market incentive to maximize the number of viewers and to reach older viewers, who are more attractive to advertisers.

The ABA also found that some programs failed to meet CTS 2(c)'s requirement that programs be "well produced using sufficient resources to ensure a high standard of script, cast, direction, editing, shooting, sound and other production elements."

The quality objective of this standard means the skillful and professional use of sufficient resources in all areas of production, from initial research of the concept, style and target audience, through the crafting and editing of the script to the final on-screen presentation. In this regard, the question of resources is no different from that of quality family or adult programs.

The ABA has found problems with scripts that were "thin, derivative, too long to engage and sustain a child audience, too focussed on adults or adult issues, or which contain gratuitous violence or unsafe practices for children," and in cast-

---

46. ABA Information Paper, Background to the CTS 2 Criteria at 2–3 (1996).
47. *Order*, 11 FCC Rcd at 10676.

ing, "with wooden performances by major characters resulting in implausible characterization."[48]

The quality of children's educational programming has likewise been a source of concern in the United States. But, the FCC identifies as a "first principle" that judgments about program quality should by made by the public, not the government.[49] It remains to be seen whether parental pressure and market forces will take care of quality concerns. If programs are low quality, children will not watch them. Because broadcasters seek the largest possible audience for advertisers, they have at least some incentive to devote resources to a quality program. However, if they do not and members of the public complain, it is unclear what action the FCC will take. It is difficult to imagine that the FCC would to refuse to count programs because of weak characterizations or slow-paced storylines, as the ABA has done. These types of judgments would be seen as interfering in the editorial discretion of broadcasters.

Although the Australians do not require programming to be educational, CTS 2(d)'s concept of "enhancing a child's understanding and experience" is similar to the FCC's broad definition of educational programming. In practice, entertainment programs with little or no educational value are not classified as C or P programs. Moreover, virtually all U.S.-made programs classified as C by the ABA would be considered educational in the United States.

Classification

Australia has also experienced controversy over who determines whether a program meets the criteria. In seeking judicial review of the 1984 CTS, broadcasters challenged the regulatory body's authority to classify a program as C before a licensee could claim credit for it. The broadcasters argued in the Federal Court that "system of prior classification amounts to the censorship of program material, a power not given in general terms to the Tribunal."[50] However, the Court rejected the argument that prior classification constituted censorship because the decision not to classify a program as C did not prevent its transmission. Moreover, the Court saw no reason to prohibit preclassification. Although striking down the pre-

---

48. Background to the CTS 2 Criteria at 5. In a recent case, the ABA found a program deficient under the quality criterion because of "weaknesses in the submitted scripts demonstrated by a confused time setting and dated storylines, poor characterisations, unrealistic language, unclear character and plot development, an adult focus, superficial treatment of themes and issues, and slow paced storylines which lack dramatic impact." ABA Press Release, No C Classification for Smiley Series, Aug. 6, 1996 (quoting ABA Chairman Peter Webb).

49. *NPRM*, 10 FCC Rcd at 6310.

50. Herald-Sun TV Pty. Limited & Orgs. v. Australian Broadcasting Tribunal, 55 ALR 53, 61, 2 FCR 24, 31 (1984).

classification requirement would leave licensees free to make their own judgments as to whether programs met the standards, the Tribunal could dispute a licensee's judgment at the next license renewal. The Court concluded that it was better for a licensee to have the certainty of knowing in advance whether his judgment shared by the Tribunal.

This ruling was reversed by the High Court, which found that while the Broadcasting and Television Act of 1942 empowered the Tribunal to determine "that only programs which meet the standards determined for children's programs should be televised within particular hours,...it does not give the Tribunal power to decide that a particular program should not be shown during those hours."[51] An Act of Parliament subsequently restored the Tribunal's authority to approve particular children's television programs as complying with the standards.

Initially, the Tribunal relied on the Children's Program Committee to review programs and scripts and make recommendations as to whether a program met the criteria. However, in March 1992, the Tribunal abolished the CPC. The Tribunal's official explanation for disbanding the CPC was that its process was too slow.[52] Yet, others have claimed that the public uproar manufactured by the broadcasters over the CPC's recommendation against classifying as C some popular children's programs led to the CPC's demise. [53]

Currently, applications are assessed initially by officers of the ABA's Children's Television subsection. They either recommend that the program be granted classification or seek further advice from specialist advisors, who include consultants from the production industry and child development experts. The final decision is made by a Member of the ABA. Classification is good for 5 years before it has to be renewed. If classification is refused, the ABA provides the applicant with a confidential statement of the reasons.

In the United States, the FCC does not assess whether a program is educational or informational in advance. Indeed, the FCC has never questioned the validity of any claim that a particular program serves children's educational needs in assessing license renewals under the CTA. Although such a challenge could have occurred under the rules adopted in 1991, in practice, it was not worthwhile for members of the public to challenge a licensee's claims when there was no requirement that the licensee air any particular amount of children's educational programming. Now that the FCC has adopted a quantitative processing guideline, dissatisfied viewers have more incentive to bring a license renewal challenge. It is likely that the FCC will at some time have to judge whether programs do or do not meet the criteria.

---

51. Herald-Sun TV Pty. Lmt. v. ABT, 156 CLR 1, 59 ALJR 514, 516 (1985).

52. ABT ANNUAL REPORT 1991–92, at 57.

53. See, for example, Patricia Edgar, *Networks Claw Back Kid's TV*, AUSTRALIAN FINANCIAL REVIEW, May 26, 1992.

For example, suppose a station applying for license renewal lists *Teenage Mutant Ninja Turtles* as one of the programs counting toward the guideline. Assume that the program meets all but the first criteria for core programming. The station claims that the program meets the first criterion, i.e., having educating children as a significant purpose, because it teaches that good triumphs over evil and includes prosocial messages about the value of friendship. If a parents' group claims that *Teenage Mutant Ninja Turtles* does not have educating or informing children as a significant purpose, how can the FCC resolve this dispute without a meaningful definition of "educational and informational"?

Even assuming that the FCC can judge whether *Teenage Mutant Ninja Turtles* has educating or informing children as a significant purpose, the timing of this decision—at license renewal—is problematic. License terms were recently extended to 8 years. A station may be unreasonably claiming programming as core programming for up to 8 years before the FCC can do anything about it. Eight years is a long time in the life of a child, and the opportunity to educate and inform many children will have been lost.

CONCLUSION

Although broadcasters in Australia have complied with the children's television quota, program producers have sometimes sought to count programming that is not specifically designed for children, is of low quality, or is not sufficiently educational or entertaining. Australia has addressed this problem by only counting programming that has been classified in advance by the regulatory authority as meeting the criteria for children's programming.

Because U.S. broadcasters face similar economic incentives, they may be tempted to claim as satisfying the 3 hour guideline programming that is not specifically designed for children, is of low quality, or is not sufficiently educational. Instead of determining in advance whether programming meets its criteria, the FCC has left that determination to licensees, while permitting members of the public to challenge the reasonableness of a licensee's claims at the end of the license term. This raises the question of whether in the absence of preclassification, a quantitative minimum will lead to the airing of quality, age-specific and entertaining children's educational programming. Because preclassification would probably not be found constitutional, the FCC must look to other means to maximize the success of the guideline. Specifically, I suggest that the FCC consider providing a more helpful definition of "educational." Second, to address quality concerns, the FCC might examine, as the Australians do, whether sufficient production resources are available for children''s programs and whether resulting production values are equivalent to adult programming. Finally, as in Australia, the FCC should review the efforts of licensees on an annual basis. Annual review would both give broadcasters a strong incentive to comply and create a body of precedent elucidating the meaning of educational and informational programming for children.

ACKNOWLEDGEMENTS

Professor Campbell teaches in a clinical program, the Institute for Public Representation (IPR), in which students provide *pro bono* legal assistance to clients on communications policy matters. IPR has filed comments in with the FCC in proceedings to implement the Children's Television Act on behalf of the Center for Media Education and others. The views presented in this chapter are solely those of Professor Campbell. She thanks the ABA staff members, particularly Liz Gilchrist, Lesley Osborne, and Debra Richards for their invaluable assistance; Professors Vicki Jackson, Patricia Aufderheide, Susan Bloch, and Dale Kunkel for their helpful comments and Gary Nelson, Louise Klees-Wallace, Amy Bushyeager, Jennifer Bier, and Erin Brown for their research assistance. A longer version of this chapter was published in volume 20 of the HASTINGS COMMUNICATIONS AND ENTERTAINMENT LAW JOURNAL.

8

AM Stereo and the "Marketplace" Decision

David W. Sosa
University of California at Davis

EXECUTIVE SUMMARY

After a long and contentious proceeding, the Federal Communications Commission (FCC) decided in 1982 to allow the marketplace to choose among competing technical standards for stereo broadcasting in the AM radio band. At the time of this marketplace decision, the FCC acknowledged the uncertain outcome of a market-based standardization process. In the case where broadcasters failed to adopt AM stereo, the FCC identified, ex ante, two possible explanations: lack of consumer demand, or market failure (based on a network externalities argument). Nevertheless, the FCC concluded that the benefits of a market-determined outcome outweighed the costs of a de jure standard. Although largely anecdotal, interpretation of the subsequent failure of AM stereo to be widely adopted, by either broadcasters or the public, has relied principally on market and nonmarket failure arguments—agency imperfections prevented the FCC from accurately perceiving the need for government intervention; and market fragmentation and consumer inertia resulted in an inefficient market equilibrium. Using station-level data from the Los Angeles, Sacramento, and San Francisco radio markets for the period 1975 to 1996, this chapter examines audience behavior as stations adopted AM stereo. Analysis of the data yields mixed results regarding the relationship between the adoption of AM stereo and audience behavior.

INTRODUCTION

Since 1934 the FCC has regulated radio broadcasting according to a public interest standard, through the allocation (what services are offered) and assignment (who is granted a license to broadcast) of electromagnetic spectrum. As part of its allocation process, the FCC long ago determined that federally mandated (de jure) technical standards for broadcasting were in the public interest. The principal motivation for this policy has been the belief that market forces cannot efficiently select technical standards.

However, in 1982 the FCC reversed 50 years of precedent, ruling "that allowing the market to determine the selection of an AM stereo system or systems...best serves consumer well-being and furthers the FCC's mandate to regulate in the public interest."[1]

In the years following the FCC's 1982 marketplace decision, AM stereo has not been widely adopted by either broadcasters or the listening public. Although there has been no reliable accounting of stereo adoption by AM licensees, I believe that present estimates of between 15% and 20% penetration nationwide are fairly accurate. Consumer adoption of receivers is likely to be lower.

The AM stereo proceeding has been widely interpreted as a combination of market and nonmarket failure. Proponents of the nonmarket (regulatory) failure interpretation argue that the FCC had sufficient information to set a de jure technical standard for AM stereo and yet failed to do so (Braun, 1994; Klopfenstein & Sedman, 1990; Meyer, 1984). Additionally, low levels of diffusion following the 1982 decision have been attributed to a failure of the marketplace to select a de facto standard, on the grounds that observed adoption patterns for AM stereo are suboptimal (Besen & Johnson, 1986; Braun, 1994; Ducey & Fratrick, 1989; Klopfenstein & Sedman, 1990). In both cases it is assumed that the selection of a de jure standard in 1982 would have been welfare improving—with a de jure standard consumers would have purchased more AM stereo receivers and more AM stations would have adopted stereophonic equipment than was observed.

This market/nonmarket failure interpretation has long been the received wisdom about AM stereo, and continues to influence debate over communications policy. AM stereo has been invoked in numerous policy debates on communications standards, including digital audio broadcasting ("DAB Standards," 1991), high-definition (advanced) television ("Broadcasters Want," 1989), and satellite television encryption ("FM America," 1986), as an example of the perils of market-based standards selection. However, most analyses to date have completely overlooked the cost-benefit analysis performed by the FCC during its decision making process on AM stereo.

In the *Report & Order* terminating the AM stereo proceeding, the FCC outlined several potential benefits from allowing the marketplace to determine an AM stereo standard. According to the FCC the benefits from allowing marketplace competition among standards include:

---

1. FCC (1982, p. 13158).

that AM stereo might fail to be widely adopted simply because consumers rejected the technology, by suggesting no observable benefits.

This chapter is organized as follows. The next section provides a brief sketch of the AM stereo proceeding and subsequent analysis. Following that, I develop a simple model of listener behavior that permits the testing of both of the FCC's concerns: market fragmentation and excess inertia. I then propose an empirical methodology, and describe the data used and the estimation technique employed. I then present the results and draw conclusions from the analysis.

AM STEREO PROCEEDINGS

The FCC was first petitioned to set a standard for stereophonic broadcasting on the AM radio band in the late 1950s, as it also considered alternative systems for stereo sound in the FM band. Radio Corporation of America, owner of the NBC network and equipment manufacturer, and Philco corporation, a consumer electronics firm, petitioned the FCC to set a standard for AM stereo in November and December 1959, respectively. Within 4 months a third petitioner, Kahn Research Laboratories (Kahn), had also filed a petition for rule making. In 1961 the FCC declined to allow a standard for AM stereo, claiming a "lack of evidence of public need or industry desire for rule changes."[7]

In early 1976 the FCC was once again petitioned to set a standard for AM stereo. Nearly 6 years later, after a tortuous and contentious proceeding, the FCC terminated the proceeding, ruling that setting a de jure standard for AM stereo would not be in the public interest. Rather, the agency concluded that setting "only those technical parameters essential to minimally acceptable stereophonic performance" was necessary. Beyond that the FCC intended for the marketplace to determine the standard, stating:

> It is recognized that allowing the market to determine the selection of an AM stereo system or systems is a bold, new step for the FCC to take. It clearly represents a change from tradition. However, it signifies a more effective and more efficient approach to achieving the public interest goals of the FCC.[8]

There were five principal groups contributing to the proceeding: proponents of the five competing AM stereo systems,[9] broadcasters,[10] manufacturers of consumer and broadcast electronic equipment,[11] intervenors commenting on an individual basis; and federal, state, and local government agencies. Of the five groups, par-

---

7. FCC (1961, p. 1616c).
8. FCC (1982, p. 13158). The 1982 ruling permitted full-time stereo (any system) on AM for the first time.

ticipation in the proceeding was dominated by system proponents, broadcasters, and (to a lesser extent) equipment manufacturers.[12]

AM stereo was touted by broadcasters, system proponents, and equipment manufacturers as an opportunity for AM stations to achieve parity with increasingly successful FM stations in terms of sound quality.[13] Nevertheless, there was considerable consensus in the docket that in spite of adding stereo capability to AM, FM would continue to be the high-fidelity medium. The principal beneficiaries of AM stereo were identified as rural audiences, unserved or underserved by FM, and automobile listeners. The automobile market was considered important because AM has more uniform coverage throughout its service area (and typically a larger service area for a given level of power), and because of certain technical considerations with respect to mobile receivers.

The AM stereo market converged on a standard within 5 years of the marketplace decision. At the time of the FCC's final ruling, the number of competing systems had been reduced to four: Harris, Kahn, Magnavox, and Motorola.[14] By the end of 1984, Magnavox and Harris had withdrawn their systems from the AM stereo market, leaving only Kahn and Motorola. On the receiver side of the market, only Motorola-compatible radios were ever produced in commercial quantities.

In 1987, the National Telecommunications and Information Agency (NTIA) released a report on AM stereo in which it declared Motorola the de facto standard

---

9. The five system proponents were Motorola, Harris Corporation, Magnavox, Belar Electronics, and Kahn Laboratories (which had formed an alliance with the Hazeltine Corporation). All five systems were capable of being received by existing monophonic AM receivers, but all systems were incompatible with one another.

10. Among the broadcasters that participated were two of the three national networks at the time (NBC and ABC), several multistation groups both large and small, independent licensees, and representative organizations (National Association of Broadcasters and the National Radio Broadcasters Association).

11. Of the equipment manufacturers, participation in the proceedings was largely by Japanese consumer electronics firms (Sony, Pioneer, Sansui, and Matsushita), U.S. broadcast electronics and semiconductor firms, and U.S. auto makers (Ford and General Motors).

12. Although in the initial phase of the proceeding (early 1977) broadcaster participation dominated all other groups, subsequent phases were characterized by increasing system proponent activity and decreasing activity from all other groups.

13. When FM stereo was authorized in 1960, AM (then referred to as "standard broadcasting") was the principal audio broadcast medium. However, by the early 1970s FM was seen as a significant threat to the commercial viability of AM. Indeed, in the three markets I examine (Los Angeles, Sacramento, and San Francisco), FM audiences surpassed AM in 1979, 1976, and 1980, respectively. See also Ditingo (1996).

14. Belar's last public activity during the proceedings was the filing of reply comments in August 1979. Braun (1994) reported that the president of Belar made his company's withdrawal from the AM stereo market official in 1981.

(NTIA, 1987). The agency reported that among AM licensees who had adopted AM stereo, over 80% used the Motorola system.

In 1992 Congress passed the AM Radio Improvement Act, which instructed the FCC to set a standard for AM stereo within 6 months (FCC, 1993a). After a brief proceeding, the FCC anointed Motorola's system as the de jure standard for AM stereo in early 1993 (FCC, 1993b). The ruling ordered AM broadcasters using any of the alternative AM stereo systems to either switch to Motorola or return to monaural broadcasting.

FRAMING THE MARKET FAILURE ARGUMENT

As was stated earlier, the most widely accepted explanation for the limited diffusion of AM stereo is that because of institutional defects within the FCC, the agency failed to establish a de jure standard, and subsequent consumer and broadcaster uncertainty regarding the outcome of a competitive selection process resulted in market failure. A direct analysis of nonmarket failure in this case is outside the scope of this chapter. Rather, I empirically test the market failure argument, nesting both of the FCC's *ex ante* explanations for a failure of AM stereo: market fragmentation and excess inertia.

Under the market fragmentation scenario, the FCC argued that "consumers are so spread out among systems that no one system has sufficient audience to survive."[15] There is no consumer inertia in this scenario, as adoption is positive for more than one system. Instead, the problem is a failure of early adopters to coordinate. The consequence is that, in later periods, consumers, for whom stand-alone benefits are small, will be unwilling to adopt because of insufficient network benefits. The presumption is that AM stereo exhibits increasing returns to adopters as a result of system scale economies, but that market fragmentation prevents the achievement of these economies.

Working within this framework, I assume that for early adopters stand-alone benefits are greater than network benefits—the "network" is simply too small in the first years of diffusion. These early adopters are the consumers who most highly value the improvement in sound quality on AM from stereophonic sound. Consequently they have the highest willingness to pay for the new service.

That market fragmentation among systems would explain adoption patterns is a testable hypothesis inasmuch as initial adopters did perceive some benefit from the new service. On the one hand, if benefits did accrue to early adopters of AM stereo, but fragmentation led to market failure, we can expect to observe a response to the introduction of AM stereo by broadcasters. On the other hand, the absence of any observable changes in AM radio audience behavior, with the introduction of stereo programming and positive adoption, would suggest that few benefits ac-

---

15. FCC (1982, p. 13158).

crued to early adopters, casting doubt, in turn, on market failure due to coordination problems and market fragmentation.

FIGURE 8.1:
Cumulative shipments of Motorola AM stereo integrated circuits:
Los Angeles, Sacramento, San Francisco.*

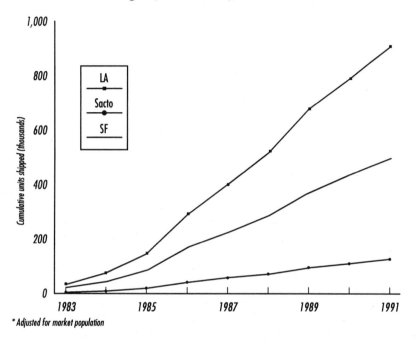

* Adjusted for market population

Under the FCC's second market failure scenario, consumer and broadcaster uncertainty result in excess inertia, as no AM stereo system is widely adopted. As
with the market fragmentation analysis, we must consider both the stand-alone
benefits and network benefits of the new service. As we did previously, we may
conclude, given positive levels of adoption among broadcasters and consumers,
that evidence of changes in audience listening patterns is essential to the validity
of an excess inertia argument. If consumers do not value the change in quality
from the introduction of AM stereo, then neither excess inertia nor market fragmentation can explain events. Rather, such evidence would suggest a marketplace
rejection of the technology.

Adoption of AM stereo by both broadcasters and consumers was positive, albeit small. By 1987 cumulative domestic shipments of C-QUAM integrated circuits, by Motorola alone,[16] totaled 11 million (see Figure 8.1). Although this
represents a small fraction of the estimated 500 million radios in the United

States, this population of receivers represents a much larger fraction of the AM radio audience. In fact, for the three radio markets examined in this study, the average audience for all AM radio stations during any 15-minute period is 5% of the market population. On the supply side, the installation of AM stereo transmitters varied widely between radio markets. For example, in 1987 adoption in the observed markets ranged from 13% of AM stations in San Francisco to 45% in Los Angeles (see Figure 8.2). Based on the given the levels of diffusion, I conclude that there has been sufficient adoption of receivers and adequate supply of AM stereo programming such that if listeners valued the improvement in sound quality in the AM band delivered by available stereo technologies there would be an observable (if small) change in audience behavior.

FIGURE 8.2:
Adoption of AM stereo by broadcasters:
Los Angeles, Sacramento, San Francisco.

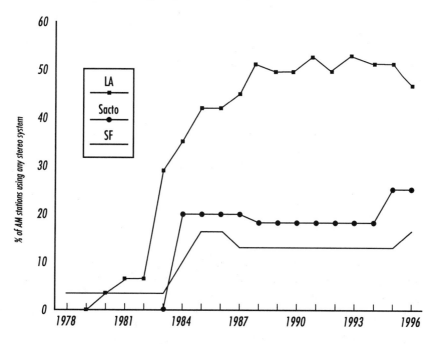

16. An integrated circuit is necessary to decode the stereo portion of an AM stereo signal. Motorola's was the only system for which these circuits were produced on a commercial scale. The NTIA (1987) estimated that 5 years after the marketplace decision, 100% of AM stereo receivers were compatible with the Motorola system.

MODELING CONSUMER RESPONSE TO AM STEREO

Consider the general case of a consumer's time allocation decision over a set of mutually exclusive alternatives—listening to one or more radio stations and engaging in other substitute non-listening activities (*e.g.,* prerecorded music, quiet, sleep, etc.). Following the standards literature, I use a strongly separable utility function to describe consumer behavior (see, *e.g.*, Katz & Shapiro, 1985). The utility consumer *i* derives from the consumption of radio services and mutually exclusive alternatives is,

$$U_i = u_i(TSL_i, x_i; a, b) + v_i(z). \qquad (1)$$

TSL_i is a vector of the time consumer *i* spends listening to each of *J* radio stations in the market. x_i is the time spent engaged in the (Hicksian) composite nonlistening activity. a is a matrix of quality parameters describing radio stations in the market (e.g., format, power), and b is a quality parameter characterizing the composite activity. $v_i(z)$ is the network benefit that accrues to consumer *i*, from joining a network of size z. As I am modeling the behavior of initial adopters, for whom the stand-alone value $u_i(\bullet)$ is superior to any network benefits, I set $v_i(z) = 0$.

Equation 1 yields specifications for audience share and intensity of listening. Station *j*'s share of the listening audience can be expressed as:

$$S_j = S(a_j; a_{-j}, b), \; j = 1, 2, \ldots J. \qquad (2)$$

The time consumer *i* spends listening to station *j* is expressed as:

$$TSL_{ij} = TSL_i(a_j; a_{-j}, b), \; j = 1, 2, \ldots J. \qquad (3)$$

In this simple hedonic formulation, demand for radio services, expressed by audience share and time spent listening, is a function of the quality of feasible listening choices and the quality of mutually exclusive non-listening activities. Whether time spent listening is positive or zero for a particular station will depend on both the quality of that station's services and the quality of other stations and of alternative activities. Thus for station *j*, own-quality, a_j, determines audience participation levels (whether or not to listen and how much to listen) given the quality of other stations, a_{-j}, and the quality of the composite activity, b. If the switch from monaural to stereo sound in the AM band represents an improvement in quality, we can expect to observe an increase in time spent listening to those AM stations that adopt an AM stereo system.

Equations 2 and 3 represent a sequential consumption decision; that is, audience decisions about whether or not to tune in to a station will affect the intensity of listening. This dependence is assumed to be one way; the intensity of listening does not have an effect on the decision of whether or not to listen.

DATA

Audience Behavior

To estimate the demand response to the adoption of stereophonic equipment by AM licensees, I collected extensive data in three California radio markets—Los Angeles, San Francisco, and Sacramento—over the period 1975 to 1996. Radio markets are defined, for the purpose of measuring audience size, by the Arbitron Company, which has been measuring audience size for commercial radio since the early 1950s. Although Arbitron measures audience size at several different market definitions, the market used in this study is the metro survey area (Metro). The Arbitron metro area generally corresponds to the federal government's metropolitan areas.

The data include measures of audience behavior during the study period, as reported by Arbitron. For commercial radio stations, Arbitron reports two station-specific audience measures: average quarter-hour persons (AQH) and cumulative persons (CUME). AQH "identifies the average number of persons estimated to have listened to a station for a minimum of five minutes during any quarter-hour in a daypart." CUME "identifies the estimated number of different people who listened to a station for a minimum of five minutes in a quarter-hour within a reported daypart. No matter how long the listening occurred, each person is counted *only once*."[17] The term *daypart* refers to the time period during which monitoring occurred. For this study the broadest measure available is used: 6 a.m. to midnight, Monday through Sunday. Additionally, the survey is limited to persons over the age of 12. Arbitron also reports estimates of the metro area population at the time of the survey. This study uses data from Arbitron's annual spring survey.

Arbitron samples from a subset of the metro population to derive its AQH and CUME estimates of audience size, using a minimum reporting standard for the data according to the following rule:

$$AQH = 0 \text{ if } AQH^* < 0.05\% \text{ of the metro population} \qquad (4)$$

$$CUME = 0 \text{ if } CUME^* < 0.495\% \text{ of the metro population} \qquad (5)$$

From AQH and CUME I calculate the market share variable (S) and the time spent listening variable (TSL). Cumulative market share for station j—the percentage of

---

17. Arbitron (1996, pp. 1–2).

the Metro population that tunes in during the daypart—is defined as;

$$S_j = \frac{\text{CUME}_j \cdot 100}{\text{metro population}}. \tag{6}$$

TSL_j is an estimate of the number of quarter-hours the average audience member spends listening to station j during the observed daypart. TSL is obtained from the following calculation:

$$TSL_j = \frac{AQH_j \cdot 504}{CUME_j}. [18] \tag{8}$$

Radio Station Characteristics

In addition to the audience estimates mentioned previously, data on various station characteristics were collected, including call letters, frequency, power, ownership, and formats broadcast. The source for these data is the *Broadcasting & Cable Yearbook* (BCY). BCY is an annual publication intended as an information clearinghouse for both broadcasters and advertisers. Individual stations report the various characteristics just mentioned annually. I assume that because broadcasters have an interest in attracting advertisers, these data have been reliably reported.

Format is difficult to identify. For example, during the study period stations in the Los Angeles market have reported over 70 distinct formats. In order to make the problem of estimating a hedonic demand system more tractable, I condense the multitude of formats into 10 broad categories: *adult*, which includes formats targeting people between the ages of 25 and 50; *rock*, representing formats intended for younger listeners; *middle-of-the-road*, which encompasses all formats intended to attract the older listener (45+); *Black*, Black-oriented formats; *country*; *talk*, which includes news, talk, and sports formats; *Spanish* language programming; *classical/jazz*; *religious*; and *other*. The various formats encompassed in these categories are described in Table 8.1.

Market Activity

During the study period, adoption behavior by commercial AM stations varies widely across markets (see Figure 8.2).[19] Sacramento was a single-system market (Motorola), but both the Kahn and Motorola systems were adopted in San Fran-

---

18. The daypart used in this study, 6 a.m to midnight, Monday to Sunday, consists of 504 quarter-hour periods.

cisco. AM licensees in the Los Angeles market adopted all of the available systems (Motorola, Kahn, Magnavox, and Harris).

TABLE 8.1
Description of Format Categories

Variable	Individual Formats
ROCK	Adult Alternative, Album Oriented Rock, Alternative Rock, Continuous Hits Rock, Classic Rock, Hits, Modern, Progressive, Rock, Top 40, Urban Rock
ADULT	Adult Rhythm, Adult Contemporary, Adult, Ballads, Contemporary, Contemporary Hits, Contemporary Soul, Love, Mellow Hits, Rock & Roll Oldies, Soft Rock
BLACK	Black, Blues, Dance, Disco, Urban Contemporary
MOR	Big Band, Beautiful Music, Easy Listening, Good Music, Instrumental, Life, Middle-of-the-Road, Nostalgia, Oldies
TALK	Business News, Information, News, News/Talk, Personality, Public Affairs, Sports, Talk
SPAN	La Mexicana, Spanish
CJ	Classical, Fine Arts, Jazz
OTHR	Arabic, Asian, Children, Chinese, Community, Culture, Diverse, Eclectic, Educational, Ethnic, Folk, International, Korean, Variety
REL	Christian, Contemporary Christian, Gospel, Inspirational, Religious
CNTRY	Country

In Sacramento, the smallest of the three markets examined, two stations adopted the Motorola system between the summer of 1983 and the spring of 1984. One station was broadcasting a news and talk format at the time of adoption, but later switched to a MOR format. The other station broadcast a country music format but switched to a talk format in 1994. Both stations continue to broadcast in stereo.

In the San Francisco market, seven AM stations adopted stereo broadcasting during the period between 1978 and 1996. Of these seven, one chose the Kahn system and the rest used the Motorola system. All seven adopters broadcast music formats. The earliest adopter was a station that initiated stereo broadcasts on an experimental basis in 1978 using the Kahn system. This station continued to employ the Kahn system until obliged by the FCC in 1993 to either adopt the Motorola system or return to monaural programming; it subsequently adopted the Motorola system.

There were two periods of adoption activity in San Francisco for the Motorola system. Between late 1983 and early 1985, three licensees adopted the Motorola system. Of these three stations, two returned to monaural broadcasting in 1993. In both cases, disadoption was coincident with a switch from music programming to a talk-based (inherently monaural) format. Between the summer of 1993 and

---

19. Among the three markets studied, the only noncommercial licensee to adopt was a religious broadcaster in Sacramento.

spring of 1996 there was renewed interest in AM stereo in the San Francisco market, as three additional stations adopted the Motorola system.

With a penetration rate of over 40% by the spring of 1985, Los Angeles had the highest incidence of broadcaster adoption among the markets studied, and anecdotal evidence suggests that the Los Angeles market may have one of the highest rates of adoption of AM stereo in the country. Although the majority of adopters initially chose the Motorola system (11 of 18 AM stations adopting stereo), three licensees adopted Kahn, three chose Harris, and one chose the Magnavox system.

The Los Angeles market further distinguishes itself from Sacramento and San Francisco in that eight stations have, at one time or another, used stereo transmission equipment to broadcast monaural programming (e.g., news and talk). Within this subset of adopters, five licensees were broadcasting monaural programming at the time of adoption.

Two AM stations in the Los Angeles market were authorized to broadcast in stereo prior to 1982, one using the Kahn system and the other Motorola. The 12 months following the FCC's 1982 decision was the most concentrated period of adoption, as seven stations began broadcasting in stereo; one using Motorola, two using Kahn, two using Harris, and one using the Magnavox system. By early 1985 three more stations had adopted, one the Harris system and the other two Motorola. In 1988 two stations adopted the Motorola system. Between 1991 and 1994 four additional stations adopted the Motorola system. During the same period four stations broadcasting in stereo returned to monaural transmissions, three of which were Spanish-language stations.

There are several interesting aspects to the history of AM stereo in the Los Angeles market. The first is the high incidence of stereo transmission equipment being used to broadcast monaural (talk) programming. The second is the pattern of multi-system adoption. By early 1984, 11 AM stations were broadcasting in stereo. However, Kahn, Harris, and Motorola had nearly equal market shares, suggesting a regional standards battle. By 1986 all but two of the stations using the Kahn, Harris, or Magnavox systems had switched to the Motorola system. Of the two remaining non-Motorola stations, one used the Kahn system until 1989 when it switched to Motorola. The other station used the Harris system, but for technical reasons its signal resembled a Motorola signal.

ESTIMATION

Equations 2 and 3 represent a sequential system. Using a semilog functional form, I estimate the following equations by the method of seemingly unrelated regressions:

$$\ln(S) = \alpha_s + \beta'_{S,V} X_{S,V} + \beta'_{S,H} X_{S,H} + \beta'_{S,MKT} X_{S,MKT} + u_s, \qquad (9)$$

and

$$\ln(TSL) = \alpha_T + \beta'_{T,V} X_{T,V} + \beta'_{T,H} X_{T,H} + \beta'_{T,AM\,stereo} X_{T,AM\,stereo} + u_t. \quad (10)$$

The semilog form is commonly used in both hedonic price estimation (Brynjolfsson & Kemerer, 1996) and discrete-choice models of product differentiation (Berry, 1994). X_v is a matrix of vertically differentiated quality variables: natural log of station power (POWER), and an indicator variable for AM or FM (BAND).[20] X_H is a matrix of horizontally differentiated quality variables: the format indicator variables ROCK, ADULT, BLACK, TALK, SPAN, CJ, OTHR, REL, and CNTRY. Because the 10 format categories used are exhaustive and mutually exclusive, the MOR variable is dropped to avoid overidentification. X_H also includes measures of format competition (AMF, FMF) and indicator variables for ownership and call letter changes (OC and CC, respectively). X_{MKT} is a matrix of market size variables (AMS, FMS, and NCS). $X_{AM\,stereo}$ is a matrix of stereo-format, cross-effects indicator variables (ROCKM, ADULTM, BLACKM, MORM, TALKM, SPANM, CJM, OTHRM, RELM, CNTRYM, ADULTK, MORK, and CNTRYK).[21] $X_{AM\,stereo}$ also includes the variable MFORM. MFORM reports, for a station using the Motorola system, the number of other stations broadcasting the same format with the Motorola system. This variable allows us to test for a critical mass effect— whether the benefits of AM stereo depend on how many stations with similar formats adopt the same technology.[22] See Annexes 8.A and 8.B for explanation of variables used in data analysis and summary statistics.

I assume that although adoption of AM stereo may affect the behavior of the inframarginal listener, as observed by a change in time spent listening, it will not affect the listener at the margin. That is, AM stereo will not affect consumers' format preferences, nor will the technology induce FM listeners to return to AM.[23] For these reasons, $X_{AM\,stereo}$ does not appear in Equation 9. I also assume that although the cost of searching, as indicated by the number of stations in the market (AMS, FMS, NCS), will have an effect on market share, it will not affect the intensity of listening. Thus X_{MKT} appears only in Equation 9.

---

20. I assume that all FM stations in the sample broadcast in stereo.

21. Because there are only four observations for the unmodified Harris system and two for the Magnavox system, and because no compatible receivers were ever commercially produced, these systems are eliminated from the empirical analysis.

22. Motorola was the only stereo system with adoption levels sufficient to test a critical mass effect.

23. The implicit assumption is that although AM stereo may represent an improvement in quality, it does not enable AM stations to achieve technical parity with FM. This assumption will bias the results in favor of finding a demand response to the introduction of AM stereo.

The standard deviations of the error terms are assumed proportional to station power. Thus for station j at year t, $\sigma_{Sjt} = (POWER_{jt})\sigma_S$ and $\sigma_{Tjt} = (POWER_{jt})\sigma_T$.

Because CUME and AQH are censored according to the rules in Equation 4 and 5, respectively, I use a two-step estimation technique first proposed by Amemiya (1974) and generalized by Lee (1978).[24] This technique assumes that the dependent variables are censored by a subset of unobservable latent variables. Whether or not CUME and AQH fall below the censoring threshold can be represented by a binary indicator variable, which is a function of the latent variables, and is estimated as a probit model.

The estimation procedure involves the following steps. First a probit regression is computed that determines the probability that the true value of the censored variable will be greater than the censoring threshold. This regression is then used to calculate the inverse Mills ratio for each radio station in each year. The inverse Mills ratio is used as an instrument for the censoring latent variables in a second-stage regression by SUR.

RESULTS

The system of Equations 9 and 10 was estimated for both an unrestricted specification and three restricted models. The restrictions imposed are coefficients on all Kahn system variables equal zero; coefficients on all Motorola system variables equal zero; and coefficients on all AM stereo system variables equal zero. Annexes 8.C and 8.D present the results for the unrestricted specification of Equation 9 and Equation 10 respectively. Due to space constraints, the results for the various restricted models are not reported. The null hypothesis being tested is that the adoption of a stereo broadcast technology by AM radio stations had no effect on the time the average listener spent with these stations. Table 8.2 presents the log-likelihood values and likelihood ratio test statistics (LR) for the various restricted models as compared with the unrestricted variant. For the Sacramento market, the null hypothesis fails to be rejected at the 15% confidence level. For the San Francisco market, however, assessment of the restricted models is not as straightforward. The specification in which the coefficients on all Motorola variables are constrained to zero fails to be rejected at the 35% significance level. Although we also fail to reject the model fully restricting all AM stereo coefficients, this comes at the lower 3.6% significance level. Finally, for the Los Angeles market, a fully restricted model and a model restricting the Motorola variables are rejected in favor of the unrestricted model. However, a model restricting the Kahn variables fails to be rejected at the 7% significance level.

---

24. See Heien and Wessells (1990) for an application to household-level food demand.

TABLE 8.2: Likelihood Ratio (LR) Tests
H_0: **Restricted AM stereo coefficients equal zero**

Model		Los Angeles Log-l	LR		Sacramento Log-l	LR	San Francisco Log-l	LR	
Unrestricted		4438			1712		4937		
Restricting:	Kahn	4434	7.15	*			4933	9.60	***
	Motorola	4418	39.44	***			4934	6.68	
	All AM stereo	4415	45.40	***	1710	5.38	4929	16.45	**

* H_0 rejected at $\alpha = 0.1$

** H_0 rejected at $\alpha = 0.05$

*** H_0 rejected at $\alpha = 0.025$

Horizontally Differentiated Quality

Because the MOR format group is dropped from Equations 9 and 10, the constant in both equations may be interpreted as an AM station broadcasting in a MOR format. Estimates of time spent listening to the average baseline station are comparable across the three markets examined (see Annex 8.D). Exponentiating the estimated intercepts suggests that on average a listener in Los Angeles spends 26.5 quarter-hours, or 6 hours and 37 minutes, per week listening to an AM MOR station ($e^{3.2773} = 26.5$). For Sacramento the estimate is 23.5 quarter-hours and for San Francisco it is 26.9 quarter-hours. Among the coefficients for other format categories, five are statistically significant in the Los Angeles market (BLACK, TALK, SPAN, CJ, and OTHR) and four in the San Francisco market (ROCK, ADULT, CJ, and REL). Generally, the results suggest that the average listener in the ROCK, ADULT, CJ, OTHR, and REL categories spends less time listening than the average MOR audience member. Conversely, listeners preferring the BLACK or TALK formats spend more time, on average, listening than do those preferring MOR. For the remaining two format types (SPAN and CNTRY), audience behavior varies across markets. Results suggest a large and positive differential in time spent listening to Spanish-language programming relative to MOR in the Los Angeles market. In all three markets, the coefficients on the measures of format competition (AMF and FMF) are small and statistically insignificant, suggesting that a measure such as the number of stations in these broad categories is insufficient to capture the effects of interformat competition on time spent listening.

Results from estimation of Equation 9 suggest that more people tune in to stations broadcasting a Rock, Adult, Black, Talk, or MOR format than to stations broadcasting in one of the other five format groupings (see Annex 8.C).

Vertically Differentiated Quality

Coefficients on the variable BAND suggest that although more people tune in to
AM radio in the Los Angeles and Sacramento markets, time spent listening to FM
is greater than to AM. In this model, station power has a strong correlation with
increased cumulative audience share in all markets. However, results suggest the
relationship between station power and time spent listening is weak.

AM Stereo

As stated earlier, only for the Los Angeles market do the results suggest that adoption of the Motorola system may have an effect on audience behavior. Although
the AM stereo coefficients are all positive and statistically significant for the Sacramento market, the likelihood ratio test fails to reject the null hypothesis that they
are jointly zero. For the San Francisco market only the Kahn system coefficients
are statistically significant. However, this result must be interpreted with caution.
Because Kahn-compatible receivers were never produced in commercial quantities, it is unlikely that this result represents a response to AM stereo, but rather
some unobserved station-specific characteristic.

In Los Angeles, a restricted model without the Kahn variables fails to be rejected in favor of the unrestricted model. Coefficients on three of the Motorola variables (ADULTM, TALKM, and SPANM) are statistically insignificant. ROCKM
and BLACKM have statistically significant, negative coefficients. These two formats typically target younger audiences and the negative coefficients suggest that
the adoption of AM stereo may have been coincident with audience migration
from AM to FM. Coefficients on the four remaining AM stereo variables (MORM,
OTHRM, RELM, and CNTRYM) are positive and statistically significant. Although these results may suggest that some consumers benefit from AM stereo,
they should be interpreted with caution. This is because the empirical model used
in this study does not fully account for the fact that broadcast licenses are hetero-
geneous properties. Station power is an incomplete characterization of the license
property. Other important factors determined by the conditions of the license, include whether or not the station uses a directional antenna, hours of operation,[25]
location on the dial,[26] and interference patterns.[27]

---

25. Whereas all FM stations are licensed to operate continuously, some AM stations may
only operate during daylight hours.

26. It is a widely held belief among radio broadcasters that a frequency assignment toward the middle of the dial is more valuable than one toward the ends, because listeners
scanning across stations will be more likely to tune into a station in the center.

27. For example, one AM station in Los Angeles experiences considerable interference
from a Mexican station on the border. An FM station in the Los Angeles market experiences
interference from a local television station.

Finally, estimates of a critical mass effect are also ambiguous. In the San Francisco market the coefficient on MFORM is negative and statistically insignificant. In the Los Angeles market the coefficient is positive and statistically insignificant.

CONCLUSIONS

These estimates of audience behavior do not provide conclusive evidence that AM stereo was rejected by consumers, but neither do the ambiguous results support the prevailing wisdom that AM stereo was not widely adopted because of market failures. The fact that by 1986 Motorola had almost 90% of the market among adopting AM stations should dampen enthusiasm for the market fragmentation rationale. Furthermore, mixed evidence of consumer response to the introduction of AM stereo calls into question the widely accepted assertion that AM stereo embodies an unambiguous improvement in sound quality.

In spite of all the hyperbole surrounding this technology, a closer examination of the AM stereo docket provides anecdotal evidence in support of the conclusions drawn from this statistical analysis of consumer behavior. First, many participants in the proceedings anticipated that although stations in major markets would adopt AM stereo, the principal beneficiaries of this new technology would be listeners in communities without adequate FM service and the mobile audience. It was widely believed that in larger markets, listeners who preferred higher sound quality would remain with FM. The second important issue was, and continues to be, the severe bandwidth constraint that AM stations face. Whereas FM stations broadcast in a 200-kHz channel, AM stations must operate within a 10-kHz channel. As the amount of information a station may transmit is a direct function of the amount of bandwidth available, AM faces a significant disadvantage relative to FM. As a consequence, there was speculation among participants to the proceeding that due to spectrum constraints, AM stations would be able to transmit very little stereo information in their signals under real world conditions. These issues raised in the docket, coupled with the results of this analysis, lead us to question the prevailing assumption that AM stereo represented an unambiguous improvement in quality.

Based on this analysis, I suggest that in the future, those interested in communications policy should be more cautious about accepting the conventional wisdom regarding the adoption of AM stereo in the United States. Although market and nonmarket imperfections may have contributed to the outcome of this adoption process, these results offer some initial suggestion that consumers' perceptions about a change in quality may tell a richer story about the failure of AM stereo to be widely adopted.

ACKNOWLEDGEMENTS

I thank Art Havenner, Tom Hazlett, Anne Selting, and an anonymous referee for helpful comments. This project was partly funded by the Program on Telecommunications Policy, University of California at Davis.

REFERENCES

Amemiya, T. (1974). Multivariate regression and simultaneous equation models when the dependent variables are truncated normal. *Econometrica, 42,* 999–1012.

Arbitron. (1996). *Arbitron radio: A guide to understanding and using radio audience estimates.* New York: Author.

Berry, S. T. (1994). Estimating discrete-choice models of product differentiation. *RAND Journal of Economics, 25,* 242–62.

Besen, S. M., & Johnson, L. L. (1986). *Compatibility standards, competition and innovation in the broadcasting industry.* Santa Monica, CA: RAND.

Braun, M. J. (1994). *AM Stereo and the FCC: Case study of a marketplace shibboleth.* Norwood, NJ: Ablex.

Broadcasters want stronger emphasis on single HDTV standard. (1989, April 17). *Communications Daily,* p. 1.

Brynjolfsson, E. & Kemerer, C. F. (1996). Network externalities in microcomputer software: An econometric analysis of the spreadsheet market. *Management Science, 42,* 1627–47.

DAB standards stalemate must be avoided, NAB's Abel says. (1991, February 20). *Communications Daily,* p.2.

David, P. A., & Greenstein, S. (1990). The economics of compatibility standards: An introduction to recent research. *Economics of Innovation and New Technology, 1,* 3–41.

Ditingo, V. M. (1996). *The remaking of radio.* Boston: Focal Press.

Ducey, R. V., & Fratrick, M. R. (1989). Broadcasting industry response to new technologies. *Journal of Media Economics, 2,* 67–86.

Federal Communications Commission. (1961). Memorandum Opinion and Order: Adoption of AM stereophonic standards not considered. *Pike & Fisher's Radio Regulation, 21,* 1616c–1616e.

Federal Communications Commission. (1982). Report and order: AM stereophonic broadcasting. *Federal Register, 47,* 13152-13167.

Federal Communications Commission. (1993a). Notice of Proposed Rulemaking: AM Stereophonic Broadcasting. *Federal Communications FCC Record, 8,* 688–92.

Federal Communications Commission. (1993b). Report and order: AM stereophonic broadcasting. *Federal Communications FCC Record, 8,* 8216–8220.

FM America weighs in: Scrambling standard sought from FCC. (1986, August 4). *Communications Daily*, 3.

Heien, D. & Wessells C. R. (1990). Demand systems estimation with microdata: A censored regression approach. *Journal of Business & Economic Statistics, 8*, 365–71.

Katz, M. L., & Shapiro C. (1985). Network externalities, competition and compatibility. *American Economic Review, 75*, 424–40.

Katz, M. L., & Shapiro C. (1994). Systems competition and network effects. *Journal of Economic Perspectives, 8*, 93–115.

Klopfenstein, B. C., & Sedman, D. (1990). Technical standards and the marketplace: The case of AM stereo. *Journal of Broadcasting & Electronic Media, 34*, 171–94.

Lee, L. F. (1978). Simultaneous equations models with discrete and censored dependent variables. In P. Manski & D. McFadden (Eds.), *Structural analysis of discrete data with econometric applications* (pp. 346–64). Cambridge, MA: MIT Press.

Meyer, J. B. (1984). The FCC and AM stereo: A deregulatory breach of duty. *University of Pennsylvania Law Review, 133*, 265–86.

National Telecommunications and Information Administration. (1987). *AM stereo and multi-system compatibility*. Washington, DC: Author.

ANNEX 8.A
Regression Variables

Variable	Description
BAND	1 if FM, 0 if AM
OC	1 if ownership change from previous year, 0 otherwise
CC	1 if change in call letters from previous year, 0 otherwise
POWER	Log of daytime station power (watts)
ROCK	1 if station broadcasts Rock format, 0 otherwise
ADULT	1 if station broadcasts Adult format, 0 otherwise
BLACK	1 if station broadcasts Black format, 0 otherwise
MOR	1 if station broadcasts Middle-of-the-Road format, 0 otherwise
TALK	1 if station broadcasts Talk format, 0 otherwise
SPAN	1 if station broadcasts Spanish language format, 0 otherwise
CJ	1 if station broadcasts Classical/Jazz format, 0 otherwise
OTHR	1 if station broadcasts Other format, 0 otherwise
REL	1 if station broadcasts Religious format, 0 otherwise
CNTRY	1 if station broadcasts Country format, 0 otherwise
AMF	Number of commercial AM stations in home market broadcasting in same format
FMF	Number of commercial FM stations in home market broadcasting in same format
AMS	Number of commercial AM stations in home market
FMS	Number of commercial FM stations in home market
NCS	Number of noncommercial stations in home market
MFORM	If AM licensee broadcasts with Motorola system, number of other commercial AM stations broadcasting in home market in same format with Motorola system, 0 otherwise
MOT	1 if AM station broadcasts with Motorola system, 0 otherwise
ROCKM	1 if Motorola stereo and Rock format, 0 otherwise
ADULTM	1 if Motorola stereo and Adult format, 0 otherwise
BLACKM	1 if Motorola stereo and Black format, 0 otherwise
MORM	1 if Motorola stereo and Middle-of-the-Road format, 0 otherwise
TALKM	1 if Motorola stereo and Talk format, 0 otherwise
SPANM	1 if Motorola stereo and Spanish format, 0 otherwise
OTHRM	1 if Motorola stereo and Other format, 0 otherwise
CJM	1 if Motorola stereo and Classical/Jazz format, 0 otherwise
RELM	1 if Motorola stereo and Religious format, 0 otherwise
CNTRYM	1 if Motorola stereo and Country format, 0 otherwise
KAHN	1 if AM station broadcasts with Kahn system, 0 otherwise
ADULTK	1 if Kahn stereo and Adult format, 0 otherwise
MORK	1 if Kahn stereo and Middle-of-the-Road format, 0 otherwise
CNTRYK	1 if Kahn stereo and Country format, 0 otherwise
IMR	Inverse Mills Ratio
AQH	Average number of persons 12 and over listening to station during a 15-minute period between 6 a.m. and midnight, Monday through Sunday
CUME	Total number of persons 12 and over listening to station for 5 minutes or more between the period 6 a.m. and midnight, Monday through Sunday
S	CUME persons as a percentage of total metro population
TSL	Number of quarter-hours average listener spends with station 6 a.m. to midnight, Monday through Sunday: TSL = (AQH*504)/CUME

ANNEX 8.B:
Descriptive Statistics

Variable	Los Angeles (n = 1408)			Sacramento (n = 511)			San Francisco (n = 1452)		
	Mean	Std. Dev.	Non-zero obs.	Mean	Std. Dev.	Non-zero obs.	Mean	Std. Dev.	Non-zero obs.
ROCK	0.169	0.375	238	0.188	0.391	96	0.165	0.371	239
ADULT	0.140	0.347	197	0.176	0.381	90	0.219	0.414	318
BLACK	0.069	0.253	97	0.018	0.132	9	0.035	0.184	51
MOR	0.173	0.378	243	0.245	0.430	125	0.218	0.413	317
TALK	0.095	0.294	134	0.078	0.269	40	0.075	0.264	109
SPAN	0.106	0.308	149	0.059	0.235	30	0.087	0.282	126
CJ	0.042	0.200	59	0.018	0.132	9	0.076	0.266	111
OTHR	0.015	0.121	21				0.017	0.128	24
REL	0.133	0.339	187	0.114	0.318	58	0.023	0.149	33
CNTRY	0.059	0.236	83	0.106	0.308	54	0.085	0.280	124
MOT	0.119	0.324	168	0.051	0.220	26	0.027	0.162	39
ROCKM	0.009	0.096	13				0.004	0.064	6
ADULTM	0.004	0.060	5						
BLACKM	0.008	0.088	11						
MORM	0.023	0.151	33	0.022	0.145	11	0.006	0.079	9
TALKM	0.039	0.194	55	0.010	0.099	5			
SPANM	0.012	0.109	17				0.002	0.045	3
CJM							0.006	0.079	9
OTHRM	0.008	0.088	11						
RELM	0.007	0.084	10						
CNTRYM	0.009	0.096	13	0.020	0.139	10	0.008	0.091	12
KAHN	0.011	0.106	16				0.010	0.101	15
ADULTK	0.004	0.060	5				0.007	0.083	10
MORK	0.006	0.075	8				0.003	0.059	5
CNTRYK	0.002	0.046	3						

ANNEX 8.C: Regression Results
Dependent Variable = ln(S)

Variable	Los Angeles Coefficient	Std. Error	Sacramento Coefficient	Std. Error	San Francisco Coefficient	Std. Error
Intercept	1.3140	0.7217 *	2.4074	0.3228 ***	1.8423	0.8310 **
BAND	-0.1761	0.0397 ***	-0.1757	0.0763 **	0.1941	0.0348 ***
OC	-0.0045	0.0588	0.0276	0.0975	0.0696	0.0544
CC	-0.3047	0.0812 ***	-0.0379	0.1055	-0.2390	0.0661 ***
POWER	0.3376	0.0107 ***	0.4006	0.0210 ***	0.4841	0.0106 ***
ROCK	0.3997	0.0788 ***	0.4015	0.1578 **	0.1062	0.0977
ADULT	0.3052	0.0759 ***	0.1177	0.1519	-0.0986	0.0667
BLACK	0.1761	0.0860 **	0.2175	0.2287	0.4900	0.1074 ***
TALK	0.3825	0.0915 ***	-0.5775	0.1484 ***	-0.1276	0.0844
SPAN	-0.1691	0.0799 **	-1.3549	0.1729 ***	-0.7135	0.0716 ***
CJ	-0.2211	0.1170 *	-0.7739	0.2569 ***	-0.0804	0.0851
OTHR	-1.2235	0.1382 ***			-0.7792	0.1229 ***
REL	-0.8456	0.0669 ***	-1.0842	0.1314 ***	-1.1796	0.1291 ***
CNTRY	-0.3445	0.0969 ***	-0.6857	0.1522 ***	-0.0571	0.0756
AMF	-0.0706	0.0110 ***	-0.1261	0.0423 ***	-0.0525	0.0102 ***
FMF	-0.0478	0.0085 ***	-0.0674	0.0263 **	-0.0124	0.0071 *
AMS	-0.0717	0.0149 ***	-0.1639	0.0346 ***	-0.1194	0.0220 ***
FMS			-0.2104	0.0221 ***	-0.0433	0.0104 ***
NCS	-0.0653	0.0233 ***	0.1328	0.0138 ***	0.0075	0.0046
IMR	0.1049	0.0025 ***	0.1135	0.0054 ***	0.0865	0.0023 ***
R^2	0.722		0.706		0.735	
Adjusted R^2	0.718		0.695		0.732	
Sample size	1408		511		1452	

* Significant at $\alpha = 0.10$

** Significant at $\alpha = 0.05$

*** Significant at $\alpha = 0.01$

ANNEX 8.D: Regression Results
Dependent Variable = ln(TSL)

	Los Angeles		Sacramento		San Francisco	
Variable	Coefficient	Std. Error	Coefficient	Std. Error	Coefficient	Std. Error
Intercept	3.2773	0.1232 ***	3.1561	0.2496 ***	3.2913	0.1155 ***
BAND	0.1513	0.0389 ***	0.1096	0.0775	0.0437	0.0346
OC	-0.0240	0.0561	-0.0096	0.0986	-0.0093	0.0539
CC	0.0029	0.0776	-0.1651	0.1067	-0.0677	0.0655
POWER	-0.0030	0.0103	0.0258	0.0213	0.0103	0.0105
ROCK	-0.1024	0.0762	-0.2534	0.1598	-0.2091	0.0967 **
ADULT	-0.0334	0.0735	-0.2160	0.1539	-0.1268	0.0661 *
BLACK	0.1473	0.0847 *	0.1313	0.2317	0.0706	0.1067
TALK	0.1674	0.0920 *	0.0300	0.1515	0.0918	0.0835
SPAN	0.2796	0.0789 ***	0.1558	0.1753	-0.0161	0.0713
CJ	-0.1874	0.1136 *	-0.1758	0.2604	-0.2227	0.0848 ***
OTHR	-0.3389	0.1419 **			-0.1429	0.1220
REL	-0.0534	0.0660	-0.0908	0.1333	-0.2155	0.1282 *
CNTRY	0.0147	0.0960	0.0232	0.1555	-0.0091	0.0754
AMF	-0.0169	0.0106	-0.0030	0.0428	-0.0066	0.0101
FMF	0.0043	0.0082	0.0267	0.0267	0.0044	0.0071
MFORM	0.0138	0.0197			-0.2546	0.1663
ROCKM	-0.2485	0.0896 ***			0.1349	0.1032
ADULTM	0.1617	0.1640				
BLACKM	-0.6419	0.1081 ***				
MORM	0.4505	0.0751 ***	0.2935	0.0828 ***	0.1825	0.1120
TALKM	0.0436	0.0986	0.2861	0.1387 **		
SPANM	0.1437	0.0902			0.2894	0.1828
CJM					-0.1048	0.0925
OTHRM	0.6738	0.1222 ***				
RELM	0.2276	0.1050 **				
CNTRYM	0.1707	0.0952 *	0.3766	0.1135 ***	0.0512	0.0747
ADULTK	0.1883	0.1434			0.2301	0.0804 ***
MORK	0.3581	0.1256 ***			0.4917	0.1124 ***
CNTRYK	-0.0335	0.1781				
IMR	0.0640	0.0024 ***	0.0802	0.0054 ***	0.0433	0.0022 ***
R^2	0.654		0.602		0.602	
Adjusted R^2	0.646		0.587		0.595	
Sample size	1408		511		1452	

* Significant at $\alpha = 0.10$

** Significant at $\alpha = 0.05$

*** Significant at $\alpha = 0.01$

IV

THE INTERNET

9

A Taxonomy of Internet Telephony Applications

David D. Clark
Massachusetts Institute of Technology

There are a broad range of objectives and opportunities that can be lumped under the heading of Internet telephony, or ITel. This chapter identifies a number of criteria that can be used to separate these different ITel applications into classes, including the degree of interoperation with the existing telephone system, and the extent to which the existing Internet must be augmented to support them. Using this framework, the chapter concludes that different ITel applications have very different motivations, and have very different implications for industry structure, economics, and regulation. The immediate opportunities for ITel involve cost reductions relative to current telephone pricing. The long-term trajectory for ITel is to become a new mode of computer-mediated human communication, which will have profound consequences for the telephone industry. This long-term form of ITel will not necessarily grow directly from the products that are being deployed now, but will come from a number of intermediate developments that can be anticipated over the next several years.

INTRODUCTION

Internet telephony, or ITel, can mean a number of different things. It can mean the use of Internet technology to replace a long distance or international provider of traditional telephone service, or an enhanced form of human-to-human communication based on the computer as the user interface, rather than the telephone. This chapter describes the rather wide range of applications that have been called ITel and tries to articulate the important differences among them, with the goal of organizing them into major classes.

The purpose of organizing ITel applications into classes is that it provides a framework around which to speculate on the broader implications of ITel. The different classes of ITel have very different justifications, and very different implications for the relevant industrial sectors involved, as well as policymakers and users.

Some ITel applications are focused on a short-term cost savings strategy that may not have strong long-term market viability. However, a possible long-term outcome of the ITel evolution is that people use computers rather than telephones to communicate. This outcome, were it to happen, could trigger a major restructuring of the telephone industry, in which separate firms provide the low-level physical connectivity and Internet service and the higher level telephone service itself.

The final speculative form of ITel just described is not practical today, because the necessary supporting features in the Internet are not in place. It is my hypothesis that ITel will evolve as a series of incremental steps. Early variants of ITel will be identified that can be deployed without first requiring as much enhancement of the Internet. These offerings will serve as experiments to prove the market, evaluate demand, explore the desirability of features, and motivate the fuller deployment of enhanced Internet service.

The chapter looks at specific examples of ITel applications in order to develop an approach to organizing and classifying them. I begin by proposing a first basis for classification, which is how much interoperation is required between the Internet and the existing public switched telephone network (PSTN). I then describe in greater detail what is implied by the long-term vision of ITel already alluded to. By listing the critical features of this long-term outcome, I develop a further checklist of features and requirements by which to organize the different ITel applications. I argue that there are ITel applications with different mixes of these features and requirements, which in turn suggests a possible evolutionary path for the future.

HOW MUCH PSTN? THE MOST IMPORTANT QUESTION

The most significant distinction between the various ITel applications is the question of how much PSTN and how much computer-based telephony is in the scheme. This chapter identifies three important classes of ITel applications.

- Class 1: Proposals with the goal of using Internet to provide plain old telephone service (POTS) between existing telephone end-user equipment. Applications of this class require technology for interconnection between the PSTN and Internet networks, but do not require access to computer-based end nodes, and can often operate across dedicated regions of the Internet.
- Class 2: Proposals that require interoperation between the existing telephone and Internet networks and provide communication between users with either computers or existing telephone sets as end nodes. This class

requires both the interoperation between Internet and PSTN, as well as the use of computer-based end nodes.

- Class 3: Proposals that use Internet-attached computers to provide some form of human communication across the packet-switched Internet. This class, the pure form of Internet-based communication, does not involve any aspect of PSTN interaction, or interworking with telephone end nodes. This class is 100% Internet and 0% PSTN.

These three classes are illustrated in Figure 9.1. This classification seems to provide the most powerful way of articulating the different broad categories of ITel.

To illustrate some of the issues, this chapter first considers an example of a Class 1 service, before looking at some of the other ways that applications can be distinguished.

EXAMPLE 1: CLASS 1—INTERNATIONAL/LONG DISTANCE POTS USING INTERNET TECHNOLOGY

This variant of ITel uses Internet technology to connect into the existing telephone infrastructure as an international or long distance carrier. Customers continue to use their local phone system and telephones, and see this just as a long distance or international alternative. Because the customers continue to use their existing handsets, the service is still essentially POTS. Although some variation in voice quality is possible, the motivation is to deliver a lower cost variant on traditional POTS telephony by using Internet technology.

FIGURE 9.1
The Three Classes of Internet Telephony

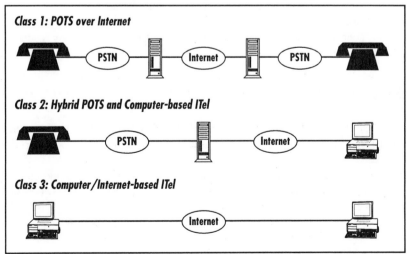

Reducing the Cost of POTS Telephony

There are three ways that a lower cost could be realized. First, some of the costs in existing telephony are artificially high, and ITel may be able to sidestep these artificial costs. Second, by efficient compression of the voice, costs could be reduced at the expense of a somewhat lower sound quality. Third, the Internet technology could deliver a lower intrinsic cost for the same service. I claim that the first factor is actually the only important one.

There has been considerable debate about the relative cost of carrying voice over the Internet and the PSTN, but there seems to be no intrinsic cost advantage to carrying POTS-style calls of the same audio quality over the two networks. The argument is that the same circuit is used in both cases, the same compression scheme could be used in both cases if the cost reduction warrants, the compression (if done) is implemented in a component of the same complexity in both cases, and both schemes can take advantage of the statistical nature of talk spurts if warranted. In the long run, the use of the Internet (or packet switching in general) does not appear to lead to greatly reduced per-minute costs for carrying a call.[1]

Although there may be few long-term intrinsic cost advantages to using Internet technology as a component of the POTS infrastructure, the current telephone prices, with regulated rates and high prices sustained by monopoly players in certain countries, appear to provide a number of options for new entrants to offer a much lower cost alternative to incumbent providers by structuring themselves as Internet providers. In the international market, these options for arbitrage are substantial. In the long distance market in the United States, prices have already been driven down by competition, so there is less advantage for arbitrage than in the international case. However, the access fees paid by traditional long distance providers to the local exchange providers currently do not apply to the Internet, so long distance provided over Internet avoids these fees. The motivation for some providers to propose long distance POTS over the Internet may be the indirect one of forcing the Federal Communications Commission to move on the resolution of the current consideration of local access charges.

EXAMPLE 2: CLASS 3—LONG-TERM INTERNET-BASED COMMUNICATION

In contrast to the application just described, which is a short-term proposition to exploit price distortions, there is a long-term vision of what Internet telephony

---

1. In the short run, the Internet solution might have lower costs, because the telephone provider must deal with the depreciation of capital equipment such as circuit switches that makes less efficient use of the same circuit. Additionally, an important second-order effect is that because an Internet infrastructure can support the signaling and operations requirements as well as carry the voice traffic, there may be some significant efficiencies in building a single Internet to support all aspects of the telephone application.

might be. The speculative end-point of ITel is a general set of applications for computer-mediated human communications. The distinguishing characteristic of this application is not lower cost, but enhanced functions. The computer and the Internet are central to this objective.

The Internet is a natural network for this application (moreso than the existing phone system) for several reasons. Packet switching allows these applications to be mixed with others over a common network. Voice can be combined with other modes of communication—text, video, shared workspace, and so on. For telephone-like applications, the Internet can deliver the signaling information all the way to the end node, so the telephone features can be implemented at the edge of the network. It will permit several calls to coexist over one physical copper pair (or other medium), and it supports advanced features such as multicast that permit many-to-many communications.

The computer provides the end node functionality. It can be used as a "call manager," keeping track of numbers and unanswered calls, assigning priority to incoming calls and redirecting them as appropriate, and logging and archiving calls. It can personalize the communications service for each user that shares the system; for example, providing a different response for business callers, friends, and strangers calling the same location. The computer can provide a sophisticated user interface to these functions (as opposed to the touch-pad interface obtained with a private branch exchange (PBX) today and can implement these functions for a single telephone line. The computer can assist in lowering costs for communications by obtaining network service from the lowest cost provider at each instant.

Differentiating Characteristics

This form of ITel differs from the Class 1 PSTN substitution described earlier in a number of important ways. These differences will turn out to be important characteristics that can be used to classify variants of ITel generally.

DIVERGENCE AWAY FROM POTS FUNCTIONALITY. Any system that includes traditional telephone handsets must interoperate in a way that is consistent with the very limited nature of that device. The richness and flexibility of the user interface envisioned by this long-term form of ITel depends on having no (or very little) need for backward compatibility with the POTS-style service. Only for the Class 3 variations (0% PSTN) does the application designer have the option of seriously diverging from the POTS-style interface.

MIGRATION OF FUNCTION TOWARD END NODE. A characteristic that is closely intertwined with this is the extent to which the location at which functions are implemented has migrated from the center of the network to the edge. The traditional telephone system has very primitive end nodes (the telephone) and much intelligence inside the network. The Internet in general represents a different balance, with intelligent end nodes (the computer) and a simple set of functions inside the switches of the network. The long-term Class 3 variant of ITel, to a large extent, will be an application implemented in the end nodes rather than inside the network.[2] Features such as call-waiting and caller ID, which today are implemented inside the telephone network, will just be implemented as software on the user's computer.

EASE OF USE. PSTN telephony is getting more difficult to use, with longer strings of numbers necessary to complete calls, prefix sequences to select long distance carriers (a concept that many consumers find very confusing), and advanced features such as call-waiting or voice mail implemented using the telephone keypad as the user interface. The addition of more features can only make this worse. On the other hand, the use of a computer as an alternative user interface might improve the situation. Additionally, if the computer were programmed to automate certain steps (such as selecting the long distance carrier based on the number being called and the current costs from different providers), increased ease of use might be combined with increased value. Ease of use is thus a factor that may get either better or worse, depending on the details of specific offerings.

AUGMENTS NEEDED TO THE INTERNET. In order to bring this Internet-based telephony application into existence, it will be necessary for the Internet to evolve in a number of ways: New protocols and technology will need to be developed and deployed. There are five significant enhancements to the Internet necessary to support Internet telephony:

QOS. The term *quality of service* (QoS) is used in Internet design to describe the ability to assign specific treatment within the network to certain flows of packets. For example, a specified flow of packets might receive an assured minimum bandwidth, or a bound on the delay of the delivery. There is work underway now in the Internet engineering and standards community to add QoS support to the Internet protocols.[3] [4]

---

2. The counterbalance to this is that for some Internet applications that have dynamic information shared among several parties, such as multiplayer games, there may be a pressure to move aspects of the application back into the core of the network. Thus, there are pressures that move function in both directions within the Internet.

3. Note that the term *QoS*, as used by Internet designers, has nothing to do with what change the user of an application may perceive as a result of using the QoS features of the network. The term QoS describes network-level mechanisms, not user-perceived quality. As a specific example, if a voice application needs a bounded delay on the packet delivery time, this can be achieved either by adding QoS mechanisms to the Internet, or by restricting the offered load (e.g., by blocking voice calls) so that the Internet infrastructure is somewhat underloaded.

Pricing for Enhanced Services. If the network infrastructure is enhanced with QoS mechanisms to provide different sorts of service, some controls will be required on the selection of these services by the users. In a private network (e.g., a corporate intranet) administrative controls may be sufficient to limit the use of enhanced QoS. However, in the public Internet, pricing seems the obvious mechanism to control the selection of enhanced QoS, as it both limits user consumption and provides increased income to compensate the provider of the service. So pricing mechanisms of some sort seem necessary as a complement to the basic QoS mechanisms.[5]

Reliability. Telephony is traditionally associated with a level of reliability greater than that of the current Internet. Providers of Internet technology (e.g., routers) may find themselves pressed to meet these greater expectations. Protocols and mechanisms for requirements such as fault detection and recovery may have to be upgraded. Providers of network infrastructure that supports Internet, such as hybrid fiber-coax cable facilities and wireless, may be similarly pressed to improve reliability and availability.

"Always on" Connectivity. Today, most residential Internet customers are not connected to the Internet at all times, but only when they explicitly dial up. To support general calling patterns among users (e.g., receiving a call without prearrangement) users would have to be connected to the Internet constantly, so they could receive a call at any time. This pattern of "always on" connection would add to the cost of providing modem-based Internet access, because it would require a modem at the central site to be provided for each user, and the modem banks provided by Internet access providers represent a significant part of their overall cost. Continuous connection would also increase the load on the telephone switches and trunks of the local exchange companies that provide the dial-up circuits. The access technology of the cable industry makes it easy to provide this "always on" service over cable, and that industry is not blind to the advantage that this offers them, in providing both ITel services and other applications.[6]

Ubiquitous Deployment. The term *network externality* is used to describe the attribute that a network is increasingly valuable to any one user as more other users

---

4. Internet services to support real-time delivery of audio and similar material are described in Wroclawski (1997) and Braden (1997).

5. A number of approaches to pricing of services in the Internet are discussed in McKnight and Bailey (1997).

6. For an alternative discussion of important enhancements to the Internet, including the "always on" feature and several other related requirements, see chapter 2 of Computer Science and Telecommunications Board (CSTB, 1996).

are attached. If only a few percent of the population are actually online, the appeal of enhanced service ITel will be minimal. General purpose ITel depends for its appeal on a sufficiently widespread deployment of the service, the Internet itself, and suitable end node equipment such as computers.

These five augments to the Internet are all required to support the general form of Internet telephony described in this example. As Table 9.1 illustrates, the short-term application of long distance POTS and the long-term application of next generation computer-based human communication differ in almost every one of these key attributes.

TABLE 9.1
Comparison of Class 1 Long Distance POTS and
Class 3 Long-Term ITel

	Application	Beyond POTS?	End-node fcn?	Ease of use	Internet Augments				
					QoS	Pricing	Reliability	Always on	Ubiquitous
1	Id POTS	No	None	Same	No	No	Local	No	No
3	Gen'l app	Yes	Major	Better?	Yes	Yes	Yes	Yes	Yes

In particular, the Class 1 variant requires fewer augments to Internet technology, because one can avoid the need for an explicit QoS mechanism in the Internet infrastructure by building a dedicated Internet used only for the long-haul telephony, and controlling the load by blocking calls. By using dedicated capacity, one also avoids the need for billing mechanisms at the Internet level. Finally, because the PSTN is being used to distribute the voice traffic to the end nodes, there is no concern with widespread deployment of the Internet or with providing the user with an "always on" service model.

There are concerns about reliability. Within the specific part of the Internet being used for carriage of voice traffic, this application does require robust equipment capable of "telephone-grade" reliable operation. However, this requirement only applies to the dedicated region of the Internet being used for voice. Because few enhancements are needed to the existing Internet to implement this application, it is feasible to undertake Class 1 ITel today, and there are commercial, Internet-based international providers of telephone preparing to offer service now.

EXAMPLE 3: TRYING CLASS 3 INTERNET TELEPHONY NOW

Internet-only telephony is actually being used today by consumers. This short-term version of Class 3 telephony represents individuals attempting to take advantage of the same price distortions as identified in the Class 1 discussion. Using only the Internet and their personal computers they establish packet flows and attempt voice communications.[7]

There is considerable skepticism that this short-term use of Class 3 ITel is commercially significant. The following issues apply:

- The use is attractive only because of current price distortions that (although entrenched) are not fundamental, and will be challenged by more organized business undertakings.
- The existing lack of support for explicit voice QoS in the Internet makes the quality of the call unpredictable. There is no evidence that the broad consumer market is interested in dealing with these fluctuations in quality.
- Because most consumers do not have a residential Internet service that allows them to be connected at all times, it requires prearrangement to receive such a call. This limits the utility of the service.

Without these Internet augments (and the others discussed earlier), this use of Internet is a bit of a "hobby" application, and there is little evidence that it will grow if all it does is emulate simple POTS calling. In the long run, a pure Class 3 offering will not survive as a simple POTS-style telephone replacement, but will be enhanced by new features that advance it away from a POTS equivalent, as described in the earlier speculation. The simple Class 3 ITel will survive only if regulators attempt to suppress the Class 1 business, in which case the Class 3 variant might persist as a consumer-activist campaign for price reform, keeping pressure on the commercial providers. It must be the case that if this use of Internet starts to become widespread, either it will be regulated to make it relatively less attractive, or other prices will shift. Therefore, it seems unlikely that this represents a use with long-term widespread benefit, so long as the functionality is simple POTS replacement.

On the other hand, because this Class 3 version of ITel more resembles what a long-term Internet telephone service might be, it can be a platform for early market entrants to position themselves as they gain experience. In the first stage of this form of ITel, the major business opportunity is providing software to the end user. The major supporting service that is required for telephony is a directory service that allows users to locate each other. Providers may position themselves to be major players in a later, more mature version of Class 3 services.

A COMPARISON OF POTS-COMPATIBLE TELEPHONY—CLASS 1. 2 AND 3

The preceding descriptions of the Class 1 and Class 3 variants of this service represent endpoints on a spectrum. There are a number of places between the two that could be realized. This intermediate, which I categorize as Class 2, uses the same gateways between Internet and PSTN as the Class 1 option, but extends the service

---

7. There are a number of vendors of PC-based ITel software today. For a collection of pointers to product information, see the web site at http://itel.mit.edu.

so that users directly connected to the Internet can interconnect to PSTN users. Computers and telephones can interoperate.

One way to assess the importance of this sort of proposal is to ask whether it adds appeal to either of the Class 1 or 3 endpoints. Looking at the Class 1 variant, with gateways connecting into the PSTN, there is little intrinsic benefit to extending this so that calls can be completed over the Internet. First, almost anyone with a computer also has a phone and can be reached without adding this option. In other words, the network externality represented by the telephone system swamps the externality represented by the Internet today. Second, there are substantial infrastructure implications if one attempts to reach all the way to the end user over the Internet, because the full range of augments to the existing Internet infrastructure to support QoS, billing, and "always on" access will eventually be required. For the pure Class 1 option, as noted already, to provide long distance or international POTS over Internet, one can build a dedicated infrastructure, which removes the necessity of these augments.

Looking instead at the Class 3 variant as a starting point, adding the ability to cross-connect to existing PSTN endpoints burdens the application with the limitation that it must interwork with POTS-style restrictions and can never evolve to new forms of service. Once that restriction is accepted, because the telephone endpoints so outnumber the Internet endpoints, the demand for the service will be generated by the telephone endpoints, which is the Class 1 situation. I therefore conclude that in the abstract, the Class 2 option of simple POTS replacement does not add much vigor to the simpler Class 1 option, and imposes a very burdensome restriction on the Class 3 option.

However, there is a specific context in which the Class 2 option has benefits. Calling *from* a computer *to* a telephone provides two specific short-term benefits. Because the charges for a long distance call are normally charged to the sender, calling *from* a computer over the Internet avoids the charges associated with the telephone system, and moves the cost into the Internet context, which is currently flat rate. At the same time, calling *to* a telephone bypasses the "always on" Internet requirement, and makes it possible to complete a call without prearrangement. Thus, calls from the Internet to a telephone have benefit in the short run. Note that this has nothing to do with the power of the computer, but only with the current costs and features in the two regimes. This Class 2 hybrid is an excellent example of a short-term opportunity with no obvious long-term utility.

The Class 2 option of POTS-style interconnection between Internet and PSTN raises interesting business questions. If the Internet were to be connected to the telephone system in a widespread way, it is not clear who would install, operate, and benefit from the gateways. They could be installed by Internet providers, by the telephone companies, or by third parties. The revenue situation is very different, depending on whether the presumed model of calling is from Internet to PSTN or the other direction, and whether the goal is to keep the call in the Internet or in the PSTN for the maximum time. The assumption in most cases is that the Internet will have a lower incremental price for a call, so the goal is to keep the call in the Internet. This implies that the existing telephone service providers will view this service with hostility, and they will need some further motivation (e.g., defensive offense) to deploy Internet phone gateways. Internet service providers (ISPs) might deploy these boxes if they can justify the cost as a part of their total service offering. This would represent a way to tie the lower level Internet service to the higher level ITel service, an example of vertical integration in the Internet industry. Third-party providers will deploy the gateways only if they can derive revenues, which implies that they must bill someone for the use; this billing will add complexity to the basic service.

FROM THE PRESENT TO THE FUTURE

The preceding examples and discussion provide a sufficient context to summarize the high-level assessment of the ITel arena. My thesis in this chapter is that the goal of the Class 1 applications is primarily a cost savings objective, which may not have strong long-term durability as a business opportunity. The Class 2 applications will evolve to provide speech access to a wider variety of network resources, and thus diverge from simple POTS. The Class 3 options represent the important long-term outcome of the ITel evolution, but the final speculative form of Class 3 ITel described earlier is not practical today, because the necessary augments to the Internet are not in place, and the utility and usability of the service features to the consumer has not been demonstrated.

It is my belief that ITel will evolve as a series of incremental steps. Early variants of ITel will be identified that can be deployed without requiring all the augments described here to be fully in place. These offerings will serve as experiments to prove the market, evaluate demand, explore the desirability of features, and motivate the fuller deployment of enhanced Internet service. The long-term options will evolve to more advanced forms, as the demand becomes clearer and the necessary augments come into place.

To further elaborate these points, this chapter describes a number of these short-term variants of ITel, to illustrate the requirements and potential of each in taking us from today to the speculative future already described.

OTHER CLASS 1 EXAMPLES

Bypass of Local Loop by Cable Providers

The goal here is to use Internet infrastructure to deliver telephone service over alternative technology (specifically cable infrastructure), bypassing the local loop and the local telephone provider.

TABLE 9.2

**Two Class 1 ITel Applications—Local Loop Bypass Using
Cable and Voice Over Private Infrastructure**

	Beyond	End-node	Ease of		Internet Augments			
Application	POTS?	fcn?	use	QoS	Pricing	Reliability	Always on	Ubiquitous
1 Loop bypass	No	Little	Same	Local	No	Local	Yes	No
3 Voice Over	No	None	Same?	Yes	No	Yes	No	No

Cable providers currently provide telephone service over cable using customer premise network interface devices dedicated to telephone service. They separately provide Internet service using so-called "cable modems" that provide high-speed packet transport over cable. If they could provide telephone service over Internet, they could provide both Internet access and telephone service using one device at the customer premise, which would appear to provide substantial cost advantages.

The cable industry is currently speculating on exactly what forms of telephony to offer over their cable plant in the short run. Offering full "first line" telephone service raises many regulatory and technical issues for that industry, such as funding of universal service and the need to engineer their infrastructure to the level of reliability of the telephone system (e.g., to remain operational when the power is off). A more advantageous short-term alternative might be to offer "second line" telephone service, which need not be as reliable or as full-functioned. This would allow them to "cream-skim" the telephone business, and steal high-profit offerings from the existing local-loop providers. For example, they could offer access to alternative long distance service and take away the resulting access charges that the local exchange carrier would receive. They can supply second lines for fax (which need not work when the power is off), and so on.

DIFFERENTIATING CHARACTERISTICS. As Table 9.2 illustrates, the only augments required for this application are to provide QoS over the cable infrastructure to mix the voice and data traffic. Cable modems currently do not support explicit QoS, but the current approaches to bandwidth allocation for cable modems could be extended to provide this support.

A very significant aspect of this variant is that while it is fully interoperable with the PSTN and uses the existing telephones of the consumer, it begins a push of function toward the edge of the network. A device would be needed at the customer premise to control the connection of the consumer's telephones and house wiring to the two external phone services—the cable and the copper loop. This box then begins to take on functions of call control, selection of provider based on cost, and so on. It represents a small step toward the migration of telephone function out of the network and into the endpoint.

Shared Use of Private Packet Networks for Voice

Corporations or other users that have procured private Internets, whether built from trunks or switched infrastructure, can carry some of their voice traffic over this infrastructure in order to make use of this investment. Because network capacity comes in large chunks (T1, DS3, OC3, etc.), there may be economies of scope that derive from combining voice and data over one infrastructure.[8] This application resembles the Class 1 long distance variant of POTS-style telephony, except that the wide-area infrastructure being used is one that is operated by a private organization, rather than as a public offering.

OTHER CLASS 2 EXAMPLES

TABLE 9.3
Three Class 2 Applications—Computer Telephony, Adding Voice to the Web, and Voice Access to Web Information

Application	Beyond POTS?	End-node fcn?	Ease of use	QoS	Internet Augments			
					Pricing	Reliability	Always on	Ubiquitous
2 Comp Tel	Slight	Some	Better?	No	No	PC: Yes	PC: Yes	No
2 Web Voice	Slight	Yes	Better?	Yes	Yes	Yes	No	No
2 Info Access	Major	No	Better?	No	No	No	No	No

Computer Telephony—Use of Computer to Control the Telephone

This application is not strictly ITel, but rather computer-mediated telephony. The concept is to connect the computer to the telephone system so that it becomes a more sophisticated user interface for advanced telephony functions. The computer could receive, process, and store voice mail; maintain a log of all incoming calls; store catalogs of called numbers; and so on. The telephone could still be used for the actual communication.

In the home, this opportunity raises several interesting and important issues. The typical home today has several phones, perhaps one in almost every room. This density of computers is not likely in the near future. So any significant use of the computer as part of residential telephony must be a hybrid that permits the telephone to be used when its convenience outweighs the primitive user interface.

In the corporate world, many employees have both a telephone and a computer. These are currently managed separately, and (for example) when an employee moves, both must be changed separately. Assuming that the employee will continue to have a computer, the opportunity here is to use the computer and its network

---

8. This opportunity applies to Internet, and also (and currently more popularly) to lower level switched technologies such as Frame Relay, which provide a slightly better control of QoS.

infrastructure as a replacement for the telephone and the PBX to which it is attached. This would reduce two systems to one, with presumed cost savings.

DIFFERENTIATING CHARACTERISTICS. As Table 9.3 summarizes, this application moves functions to the edge of the network (a computer or similar consumer device) while continuing to interwork with the existing PSTN and the resulting POTS-style service. It implies high reliability of the computers and networks, and computers that are always on and available to process calls.

Adding Voice to the Web

One of the current emerging examples of voice as a component of multimedia computer application is adding voice communication to Web pages. The concept is that a user browsing a Web page can click a button and talk to a representative of the company providing the Web page, thus merging the Web with 800 numbers.

This is described as a Class 2 application because the client side is a computer, and the current implementation at the server side is to connect the incoming call into the existing call dispatching equipment that deals with PSTN calls. This is a rather powerful hybrid, because it could be possible for the representative receiving the call to have access to the computer information that the customer is seeing. So this option mixes POTS-style telephony and computer-based multimedia functions.

This application, if mature, will shift 800 traffic onto the Internet, which could adversely impact revenues of the existing telephone providers. The voice calls are carried as far as possible across the Internet, and only connected into the existing telephone system at the premise of the Web provider.

DIFFERENTIATING CHARACTERISTICS. Because this application carries the voice across the Internet as far as possible, connecting into the existing telephone system only at the premise of the server, several of the Internet augments discussed earlier will be required to make this service real. Wide-scale introduction of QoS and the related pricing mechanisms will be required, but it does not require "always on" operation (the consumer originates the calls), nor does it depend strongly on ubiquitous Internet deployment. Any consumer with the service can fully benefit from it.

Voice Access to Information on the Web

In the long run, the future of the Class 2 hybrid (PSTN/Internet interconnection) is not to provide simple voice communication between humans, but to provide voice access, within the POTS paradigm, to a range of new Internet-based services. The use of touch-tone selection to navigate services is a primitive example of this, but the mature form will involve computerized voice understanding and conversation between a human on a telephone and a computer, which then reaches out into the Internet to obtain services and information for the user.

Services such as this exploit telephone calls across the PSTN, but do not overall much resemble a classical phone call, as the device at the other end is not a human but a computer, and the goal of the call is to obtain access to information located within the Internet.

DIFFERENTIATING CHARACTERISTICS. This service requires no augments to the Internet, as the voice only passes over the PSTN. These sorts of applications can be deployed as soon as the speech understanding issues have been resolved.[9]

CLASS 3 EXAMPLES

TABLE 9.4
Three Class 3 Applications—Teleconferencing, Consumer Multi-Person Applications, and Telework

Application	Beyond POTS?	End-node fcn?	Ease of use	QoS	Internet Augments			
					Pricing	Reliability	Always on	Ubiquitous
3 Teleconf	Yes	Major	Better?	??	No	Yes	No	No
3 Chat, etc.	Yes	Major	Better?	Yes	Yes	No	No	No
3 Telework	Yes	Major	Better?	Yes	Yes	Yes	Yes	Yes

As discussed earlier, the long-term form of Class 3 Internet-based telephony involves a major migration away from the assumptions of the traditional telephone system, and requires a number of augments to the current Internet. However, there are a number of Class 3 applications that are more limited in the objectives and thus in the augments that they require; see Table 9.4. These represent first steps that the industry will take, as it explores the space of real Internet-based voice communication.

Teleconferencing

This application exploits the power of the computer to move beyond simple POTS in support of human communications. The objective is to augment simple voice communications with other modes such as video, multiway (many-to-many) communication patterns, and shared workspace and other groupware tools.

---

9. Early examples of systems that employ computer-based voice understanding are available in the research lab today. For example, the Jupiter system developed by the Spoken Language Systems group at the MIT Lab for Computer Science allows a user to call and have a phone-based conversation with a computer concerning the weather, based on current weather information that the computer retrieves from the Web.

DIFFERENTIATING CHARACTERISTICS. There are several patterns of teleconferencing. One is within the corporate campus. Achieving the necessary QoS might be achieved by overprovisioning in the short term, which will permit this application to be initiated without explicit QoS support in the Internet routers. The second pattern is among the sites of a corporation. This will require support from the Internet technology similar to that discussed earlier.

The third mode of teleconferencing is telework, which is sufficiently different that it is discussed separately later.

An important aspect of Internet teleconferencing is the many-to-many mode, which is more natural in the Internet, but requires special arrangement (a conference call) in the telephone system. Many-to-many communications greatly benefits from Internet multicast, which is only now being deployed, and still suffers from concerns about scale and robustness.

Chat Room, Games, and Other Online Real-Time Applications

Chat rooms and open group interaction share with teleconferencing the basic objective of linking together a number of people in a common real time experience. However, the details are very different.

First, this application is a leisure activity, and thus targets the consumer at home, not the worker at a place of business. Second, the groupware components might be very different. Third, the open nature of the group will call for a more complex directory and group location service.

This could be considered in the context of a hybrid Internet/PSTN mode (Class 2), but is more likely to succeed as a pure Class 3 application, where the groupware modalities can be exploited to enhance the experience. The possibility of some user having multimodal participation while others have only voice seems unappealing.

DIFFERENTIATING CHARACTERISTICS. This application requires QoS, pricing, reliability, and reasonable overall bandwidth to the consumer at the residence.

Telework

Telework is a variant of teleconferencing carried out from the home or other remote location.[10] The objective here is to make teleconferencing available at the residence. Whereas corporations can deploy teleconferencing within and among their business sites, it will be the public ISPs that provide (at least the

---

10. More generally, telework could imply only more primitive network access for shared files, but we choose to use the term here to capture the human-to-human communications aspects of remote work.

infrastructure for) telework. The bandwidth and service requirements for telework may somewhat resemble the requirements for chat rooms and other recreational applications. This application will thus provide an opportunity for the consumer-oriented ISPs to play in the business sector and diversify their revenue base.

DIFFERENTIATING CHARACTERISTICS. Most corporations seem to desire to roll out a telework option only when the infrastructure is widely in place. Thus, there is a requirement for widespread deployment of Internet access.

CONCLUSIONS-THE LONG-TERM IMPLICATIONS OF ITEL

TABLE 9.5
Summary of ITel Applications

	Application	Beyond POTS?	End-node fcn?	Ease of use	QoS	Internet Augments Pricing	Reliability	Always on	Ubiquitous
1	Id POTS	No	None	Same	No	No	Local	No	No
1	Loop Byps	No	Little	Same	Local	No	Local	Yes	No
1	Voice Over	No	None	Same?	Yes	No	Yes	No	No
2	Comp Tel	Slight	Some	Better?	No	No	PC: Yes	PC: Yes	No
2	Web Voice	Slight	Yes	Better?	Yes	Yes	Yes	No	No
2	Info Access	Major	No	Better?	No	No	No	No	No
3	Teleconf	Yes	Major	Better?	??	No	Yes	No	No
3	Chat, etc.	Yes	Major	Better?	Yes	Yes	No	No	No
3	Telework	Yes	Major	Better?	Yes	Yes	Yes	Yes	Yes
3	Gen'l App	Yes	Major	Better?	Yes	Yes	Yes	Yes	Yes

Implications for Timing

My hypothesis is that ITel will evolve from early offerings that can be assembled today into mature forms that have diverged substantially from simple POTS-style telephony. Looking at Table 9.5, some options require fewer Internet augments to deploy, such as teleconferencing, voice access to information, and computer-mediated telephony. These can happen sooner, and will thus serve to explore the demand and position players in the market. Teleconferencing, for example, will explore the utility of other modes of communication, such as video and shared workspace. The end node device supporting the user, whether PC or specialized server, will start to implement a sophisticated user interface to ITel services, and will start to act as the user's agent in implementing key ITel functions. In parallel with these first steps, ISPs will start to implement augments such as QoS and pricing. These will permit the next applications, such as voice over the Web and leisure activities such as games and chat rooms. With sufficient success in the market-place, there will be enough penetration of Internet service to make the final form of Class 3 computer-mediated communication practical.

Implications for Industry Structure

The speculative final form of computer-mediated human communications implies a substantial change in the structure of the telephone service and the industry that provides it. The design of the Internet, with open interfaces between different service layers, tends to create an industry structure in which the lower level service providers (the ISPs) do not have a substantial competitive advantage in supplying higher level services such as Web hosting.[11] This pattern differs from that of the telephone industry, where the provision of physical facilities has been linked to the higher level telephone service. Despite that history, if telephony (or more general-ly, Class 3 computer-mediated human communication) were to move to the Inter-net, there is no reason why it would be immune to this separation of lower and higher level services among different players. In that case, local exchange carriers would be shifted into a role where they provide copper loops, and perhaps provide Internet service over those loops, but have no consequential advantage in provid-ing ITel. The ISP participates in the implementation of the higher level ITel service only to the extent that it provides a selection of QoS and a set of supporting ser-vices that exist in the network.

In fact, ITel ceases to be a single unified service at all. As the end node com-puter, owned and provided by the subscriber, becomes part of the telephone con-text, it will accentuate the fracturing of the service. The implementation of key functions at the end node removes them from the control of the telephone indus-try. ITel becomes a service built out of software purchased by the consumer, op-erating over a general (e.g., Internet) communications infrastructure that is independent of applications.

Another aspect of the fragmentation of the industry is the manner in which sec-ondary services are supported. The services that have been mentioned here include directory services such as white and yellow pages. Yellow pages, which are essen-tially advertising, can be expected to become a very competitive business, the final form of which is hard to predict. Already, however, Web-based services are begin-ning to appear, and there is no longer any telephone company monopoly on direc-tory services.

---

11. For a discussion of the overall design of the Internet, and the role of open interfaces in fragmenting the market segments, see chapter 2 of CSTB (1994). For a discussion of some of the specific business consequences, see CSTB (1996).

Policy Considerations

Faced with the prospect of the vertical disintegration of the telephone industry, the incumbent players may reasonably try to create circumstances that forestall or mitigate this potential. One leading indicator for the long-term form of ITel is the standards and interfaces that are proposed by major industry players. Several examples are obvious. For example, directory services are just emerging, and there is scope for these standards to either encourage or discourage the interworking of different services as a uniform overall directory system.

The form of ITel described here, with major functions implemented in the end nodes, can only be realized if the necessary control information (what the telephone industry calls *signaling*) is carried end to end across the Internet. This is the natural mode for Internet control protocols, but one could imagine an attempt to close or restrict these protocols in some way. If, for example, the called end node could not determine the number and identity of the calling party, this would help maintain ITel as a centralized application. If a Class 2 hybrid form of ITel does succeed in the market, the interconnection of the telephone and Internet telephone service will require some interface to telephone signaling protocols. The two extremes are use of touch-tone signals, or the internal signaling protocols of the telephone system, Signaling System 7, or SS7. These would represent very different modes of access into the existing functions of the PSTN, and might cause ITel signaling to grow in different ways.

Regulation of ITel has been proposed, and may again be proposed in the future, based on the observation that in some forms it is similar to the service of the PSTN.[12] The hypothesis here is that most of the Class 1 alternatives in fact exist to exploit pricing distortions, and thus arguments about regulation are germane to Class 1 ITel. However, if the hypothesis is true that the final form of ITel will result from the evolution of Class 2 and Class 3 applications, it seems more difficult to apply a similar service criterion. The Class 3 application will diverge greatly from the function of existing telephony. The Class 2 applications will continue to exploit a "phone call" to access the Internet, but if the person on the telephone is talking to a computer that is reading a Web page as the overall service, it seems difficult to describe the collective event as resembling POTS. This suggests that any attempts to regulate ITel, based on looking at the Class 1 variants, will quickly unravel in definitional confusion.

One of the consequences of this predicted evolution may be a gradual loss of reliability, dependability, and utility of the telephone service. This trend is already visible as a consequence of divestiture, where some people are already so confused that they are unable to make long distance calls from pay phones, rogue pay phones lure the unwary, bills are unpredictable, there is no accountability for dis-

---

12. See Werbach (1997) for a recent discussion of options for regulation of the Internet.

rupted service, and so on. There could be a possible consumer backlash that demands a more predictable, reliable, and usable service. One outcome could be regulation as to the characteristics of a future telephone service, regulation more akin to safety and consumer protection than pricing and cost recovery.

As the industry becomes more and more fragmented, it will become less and less easy for the various industry players to meet and agree on common objectives and approaches. This will contribute to a further deterioration in the overall utility of the service, and a resulting pressure for forces other than those of autonomous private industry to come to bear on the problem. These issues will have to be addressed and resolved in a multinational context.

ACKNOWLEDGMENTS

I gratefully acknowledge the comments and suggestions of Sharon Gillett, Andrew Sears, Russ Newman, Bill Lehr, John Wroclawski, and Lee McKnight. This work was funded by the MIT Internet Telephony Consortium and its industrial partners.

REFERENCES

Braden, R. (Ed.). (1997). Resource ReSerVation Protocol (RSVP)–Version 1 Functional Specification. Internet RFC 2205. Available online at ftp://ds.internic.net/rfc/rfc2205.txt.

Computer Science and Telecommunications Board. (1994). *Realizing the future: The Internet and beyond.* Washington, DC: National Academy Press.

Computer Science and Telecommunications Board. (1996). *The unpredictable certainty: Information infrastructure through 2000.* Washington, DC: National Academy Press.

McKnight, L., & Bailey, J. (Eds.). (1997) *Internet economics.* Cambridge, MA: MIT Press.

Werbach, K. (1997). *Digital tornado: The Internet and telecommunications policy.* OPP Working Paper 29, Washington, DC: Federal Communications Commission.

Wroclawski, J. (1997). Specification of the controlled-load network element service. Internet RFC 2211. Available online at ftp://ds.internic.net/rfc/rfc2211.txt.

10

Internet Telephony:
Costs, Pricing, and Policy

Lee W. McKnight
Brett K. Leida
Massachusetts Institute of Technology

INTRODUCTION

This chapter presents a cost model of Internet service providers and Internet telephony and assesses its business and policy implications.[1] The term *Internet telephony* has been broadly applied to a family of applications that typically includes (real-time) voice communication, at least partially over a network using Internet protocols. This is in contrast to traditional telephony or plain old telephone service (POTS), which occurs solely over a circuit-switched telephone network. However, distinctions between Internet telephony and traditional telephony become less clear when one considers telephony services that bridge packet-switched and circuit-switched networks. In Chapter 9 of this volume,

1. Support for this research from the MIT Internet Telephony Consortium (http://itel.mit.edu) is gratefully acknowledged. This chapter is based on Brett Leida's SM thesis entitled "A Cost Model of Internet Service Providers: Implications for Internet Telephony and Yield Management." We would like to thank ITC staff members and companies for providing valuable feedback via five conference calls in March and April 1997, ITC meetings in May 1996, and January and June 1997. We would also like to credit NMIS for partial support, NSF Grant NCR-9307548, and the students of the MIT Telecommunications Modeling and Policy Analysis Seminar (TPP91) for an initial model that was presented to the Federal Communications Commission in May 1996 (Students, 1996). This chapter extends slightly the model of Leida (1998) to explore the impact of Federal Communications Commission policy decisions on costs.

Clark classifies various types of telephony services that can be realized using these once disparate networks. The type of Internet telephony analyzed in this model is what Clark calls Class 3 Internet telephony: computer-to-computer Internet telephony in which two computers communicate over the Internet via a modem connection or a direct network connection.[2]

It is shown that a moderate use of computer-to-computer Internet telephony can raise the costs of an Internet service provider (ISP) by as much as 50%. Pricing and policy issues arising from Internet telephony services are also briefly addressed in this chapter. We conclude that both new pricing strategies and supportive policy frameworks are needed for Internet telephony services to recover costs and to integrate the Internet and telecommunications industries.

COST MODEL OF ISPs

A cost model of ISPs has been developed by the MIT Internet Telephony Consortium (ITC)[3] and the MIT Research Program on Communications Policy (RPCP). The model quantifies the impact on an ISP's costs due to an increased use of Internet telephony. Two scenarios are modeled: a baseline scenario representing current ISPs in which the principal use of the network is for Web browsing and there is essentially no Internet telephony; and an Internet telephony (IT) scenario in which the ISP sees a substantial increase in use of computer-to-computer Internet telephony by its subscribers.[4]

The model is used to identify the costs of end-to-end Internet service for various types of users (dial-in, leased-line, etc.). These costs are broken down into five categories:
- *Capital equipment*—the hardware and software of the network
- *Transport*—the leased lines of the network and interconnection costs
- *Customer service*—staff and facilities for supporting the customers
- *Operations*—billing, equipment and facilities maintenance, and operations personnel
- *Other expenses*—sales, marketing, general, and administrative.

The next section describes the fundamental assumptions in the model, and the following section describes the elements included in each of the five cost categories.

---

2. Ongoing work within the MIT Internet Telephony Consortium models Internet telephony across gateways, both within corporate intranets and over extranets. Results of this work, however, are not available as of the date of this writing.

3. More information about the ITC can be found at http://itel.mit.edu/.

4. As noted in footnote 2, the issue of gateway traffic (i.e., phone–gateway–phone) is not modeled here. Other work in the ITC is analyzing gateway issues.

Key Model Assumptions

The ISP cost model presented here has over 300 input parameters. Given limited space, each parameter cannot be explained in this chapter. However, the key assumptions are presented in the following.[5]

FIGURE 10.1
Hypothetical ISP network architecture

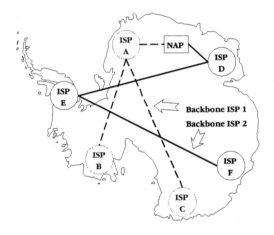

Model represents an Access and Backbone ISP with Leased Transport

ISPs come in all shapes and sizes. Firms with international, facilities-based networks are called ISPs, as are firms with a few modem racks and a leased line. A model that captures the costs for both types of firms must be carefully designed.

Figure 10.1 shows a hypothetical market where there are two principal types of ISPs: backbone ISPs (1 and 2 in the figure) and access ISPs (A through F). Each access ISP has a backbone ISP that connects it to other access ISPs. The backbone ISPs interconnect at network access points (NAPs). The ISP-entity represented in the cost model is a single backbone ISP and its associated access ISPs. Hence, the model results do not necessarily correspond to the costs of any particular ISP given the current (U.S.) market structure of separate backbone and access ISPs. However, the model results do represent the total costs of providing end-to-end Internet service, which is the intended goal of the model.

---

5. For a more detailed description of each model parameter see Leida (1998).

FIGURE 10.2
ISP topology

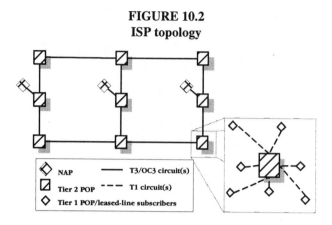

NAP — T3/OC3 circuit(s)
Tier 2 POP --- T1 circuit(s)
Tier 1 POP/leased-line subscribers

The network of the modeled ISP is shown in Figure 10.2. The backbone links are connected at nodes called Tier 2 points-of-presence (POPs) and the access nodes, Tier 1 POPs, are connected in a star network to a Tier 2 POP.[6]

- **Demand is approximated by five subscriber classes.** It is further assumed that identifying costs for the following types of subscribers captures sufficiently the costs of an ISP: residential dial-in subscriber, business dial-in subscriber, ISDN (128 kb dial-in) subscriber, 56 kb (leased-line) subscriber and T1 (leased-line) subscriber.[7]

- **Subscriber prices and levels are from the U.S. market.** Statistics from 1996/97 subscribers for the various services in the United States were used to determine revenue for the modeled ISP.[8]

- **Bandwidth per dial-in user is 5 kbps for baseline scenario and three times that for IT.** It is assumed that Internet telephony is used 33% of the time in the IT scenario. The result is that per-user bandwidth increases by a factor of 1.66 in the IT scenario.[9] Further, it is assumed that dial-in subscribers will increase their holding time and call arrival rates each by 20% in the IT scenario.[10]

---

6. Further discussion about ISP architecture is found in Leida (1998).

7. The sole distinction between residential and business dial-in subscribers is varying access patterns. Residential subscribers are assumed to request access primarily in the evening and business subscribers during the day. See Leida (1998) for more details on access patterns.

8. Sources include: FIND/SVP (1996); Forrester Research (1996); Morgan Stanley (1996); and *Boardwatch*, (1997).

9. $1.66 = 300\% * 33\% + 100\% * 67\%$.

10. Because the leased-line subscribers' users are "always on" the network, the concept of dialing in does not apply to them.

Principal Cost Categories

The ISP's costs are separated into five principal categories: capital equipment, transport, customer service, operations, and other expenses (which include sales/marketing and general/administrative).

Each cost element (e.g., router, billing, or marketing costs) is determined based on assumptions about how large the costs would be for the given number of subscribers. Once the total cost of an element is known, its cost is allocated to each type of subscriber based on the relative amount of use by each type of subscriber. Carrying out similar calculations for each cost component permits the model to determine the cost per subscriber for each type of subscriber.

CAPITAL EQUIPMENT. Capital equipment includes that which is found in the Tier 1 and Tier 2 POPs. Figure 10.3 and Figure 10.4 show how the ISP capital equipment[11] is interconnected at a (Tier 1 and Tier 2) POP. Capital investments are converted from a one-time, fixed cost to a leveled, annual cost by using a cost of capital rate.[12]

FIGURE 10.3
Tier 2 POP

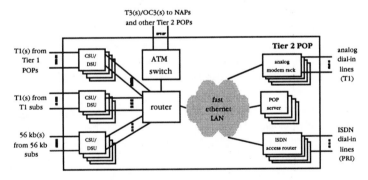

Each piece of capital equipment is sized based on assumptions of users' access patterns and bandwidth requirements. Once the total cost of a piece of equipment is known, its cost is allocated to the various types of subscribers based on the rel-

---

11. ISP capital equipment considered in the model are: analog modems, content housing server, 56 kb CSU/DSU, Cisco 7513 serial port card, Fore 4 port DS3 card, ISDN access router, LAN-10 Mbps ethernet, T1 CSU/DSU, Cisco 7513 ATM card, Fore 4 port OC-3 card, POP server, LAN-100Mbps ethernet, Cisco 2500 router, Cisco 7513 router chassis, and a Fore ASX-200BX ATM switch chassis.

12. See Leida (1998) for details.

ative amount of use by each type of subscriber. For example, analog modems are sized based on how often the residential and business dial-in subscribers call and how long they stay connected. Once the required number of modems is known (and hence, the cost), the cost is split between the residential and business dial-in subscribers based on how much traffic each type of subscriber generates. For this particular piece of capital equipment, no modem costs are allocated to the ISDN, 56 kb, and T1 subscribers because they do not use this type of equipment. Similar calculations are carried out for each piece of capital equipment.

TRANSPORT. The transport costs of the ISP are comprised of costs due to leased lines to connect the Tier 1 and Tier 2 POPs (T1s, T3s, and OC-3s) and costs due to incoming analog and ISDN phone lines (T1 and PRI) to connect the dial-in subscribers. In addition, monthly costs for the ISP to interconnect at a NAP are included in transport costs.

The costs for the leased lines are based on published tariffs by telecommunications providers such as AT&T, MCI, and Sprint. Many telecommunications providers offer substantial discounts for a customer, such as an ISP, who leases circuits in bulk. If the customer commits to a certain number of dollars per month, it will receive a discount according to the range in which its commitment falls. Such bulk discounts are taken into account in the model and are based on published figures by the previously mentioned carriers.

As with the capital equipment costs, transport costs are allocated to the various types of subscribers (e.g., ISDN PRI costs are allocated entirely to the ISDN subscribers), whereas the backbone costs are allocated to each type of subscriber according to the relative amount of bandwidth required for each type of subscriber.

CUSTOMER SERVICE. Customer service is furnished by representatives who provide technical support via the telephone to the subscribers. It is assumed that all dial-in subscribers (analog and ISDN) will be making calls to customer service. Additionally, technical representatives of subscribers with leased lines, (i.e., 56 kb and T1 subscribers) will also call customer service, but it is assumed that these subscribers will have their own internal end user support, so that the end users are not calling the ISP's customer service.

For the model, the perspective is taken that customer service is outsourced by the ISP. Hence, instead of determining how large a staff is needed, one determines how many call-minutes there are and what is the cost per minute charged to the ISP.

OPERATIONS. Operations correspond to the routine tasks necessary to keep the ISP functioning. Operations costs fall into three principal sections: network operations and maintenance, facilities, and billing.

Network operations and maintenance costs include those for maintaining the hardware and software of the network, as well as the personnel needed to carry out these responsibilities. The costs for the maintenance are based on a percentage of the total costs for the capital equipment, and the personnel costs are based on the number of people needed to maintain the given number of POPs.

Facilities costs are those associated with maintaining a physical space for each POP. Included are such costs as building rent, electricity, heat, and so on. The costs are based on an expenditure in dollars per month for each type of POP.

The costs of billing for Internet service include those of rendering a monthly bill. There is a fixed fee for generating each bill, and each subscriber receives one bill per month.

OTHER EXPENSES. The remaining costs seen by an ISP are included in an Other Expenses category. These costs include sales/marketing and general/administrative. Although these costs are not the focus of this study, they are nonetheless part of an ISP's costs and are included to provide a perspective relative to the other cost categories.

TABLE 10.1
Baseline Scenario Cost Summary (Leida, 1998)

Category	Cost	Distribution
Capital Equipment	$3,349,000	11%
Transport	$7,242,000	24%
Customer Service	$7,927,000	26%
Operations	$3,445,000	11%
Sales, Marketing, G & A	$8,725,000	28%
Total Monthly Cost	$30,688,000	

Sales and marketing costs are those used to attract and retain subscribers. These costs are based on a percentage of revenue. The value for the percentage is based on figures taken from annual reports of ISPs and other telecommunications service providers.

General and administrative (G & A) expenses consist primarily of salaries and occupancy costs for administrative, executive, legal, accounting, and finance personnel. Similar to sales and marketing costs, G & A costs are based on a percentage of total costs. The value for this percentage is also derived from annual reports of ISPs.

COST MODEL ANALYSIS AND INTERPRETATION

This section presents the results of the cost model for the baseline and IT scenarios. First, results are presented for the baseline scenario, and then results are presented for the IT scenario. Sensitivity analysis is discussed and the results are summarized.

TABLE 10.2
Revenue and Cost Comparison

	Magnitude	Residential	Business	ISDN	56 kb	T1
Revenue	$26.7 M	69.9%	0.3%	1.3%	2.7%	25.8%
Cost	$30.7 M	69.3%	0.3%	1.0%	2.2%	27.2%

Baseline Scenario Results

As described previously, the baseline scenario represents an ISP whose users are primarily browsing the Web. This scenario is intended to represent an ISP in the 1997 time period.

Table 10.2 shows the cost results for the baseline scenario. An initial conclusion is that the cost is slightly greater than the revenue. This is not necessarily surprising because many ISPs have had difficulty operating profitably.

The cost and revenue distribution across the subscriber base indicates that no type of subscriber is being substantially subsidized.[13] This indicates that the market for providing Internet access services is relatively efficient and competitive.

Table 10.1 shows the results across the various cost categories. Capital equipment and operations costs each represent approximately half the costs of the other three categories. In general, however, no particular cost category dominates the ISP's costs. Taken from another perspective, all cost categories play an important role in determining an ISP's costs.

FIGURE 10.4
Subscriber cost distribution for baseline scenario

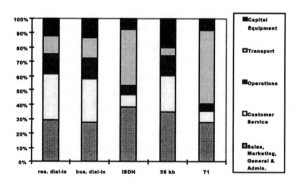

---

13. This is in contrast to the PSTN, for example, where, based on government desires, the business subscribers subsidize the residential subscribers.

Figure 10.4 shows the cost distribution for each type of subscriber. Note that this distribution varies substantially among subscriber type. For example, transport represents only a small portion of cost for dial-in subscribers but a large portion of cost for T1 subscribers. Similarly, customer service is a larger share of dial-in subscribers' cost than T1 subscribers' cost.

The baseline scenario analysis yields the following conclusions:

1. No particular cost category dominates the ISP's costs.[14]
2. There is a substantial variation in cost distribution between the different types of subscribers.
3. The ISP's total cost distribution will vary with the subscriber mix and the individual cost distribution.[15]
4. Nontechnical components represent a substantial portion of each subscriber type's costs.
5. ISPs are losing money.

IT Scenario Results

In this section results are presented for the IT scenario. First the total ISP costs are presented and then individual subscriber costs for dial-in and T1 subscribers are presented.

Comparing the IT scenario to the baseline scenario, costs in all categories increase in the IT scenario; however, some categories are affected more than others are. The bottom line for an ISP is that *revenues will increase slightly, while costs will increase substantially with only a moderate use of IT.* Hence, ISPs need to consider how to minimize the cost impact of IT and/or how to recover additional revenue if they hope to operate at profitable levels. The comparative results for the baseline and IT scenarios are shown in Figure 10.5.

At 28% of total costs, transport costs become the largest cost category in the IT scenario. The implication for ISPs based on this result is that an ISP that operates its network most efficiently will have a competitive advantage over other ISPs if the IT scenario takes place. Such efficiencies could come from scale economies, facilities-based networks, or network optimization techniques. However, if one believes that the market for transport is already efficient and that transport is essentially a commodity, then there would be fewer opportunities for competitive advantage resulting from owning a network. Even so, network optimization techniques would prove advantageous whether or not the ISP owns or leases its network.

---

14. However, this is only true for the mix of subscribers used in the baseline scenario. Other mixes of subscribers would yield different results.

15. Hence, if the subscriber mix changed, the ISP's cost distribution would be weighted by the number of each type of subscriber. For example, if the ISP had only T1 subscribers, its cost would be distributed just as the T1 subscribers' cost is distributed.

FIGURE 10.5
Comparative cost results (Leida, 1998)

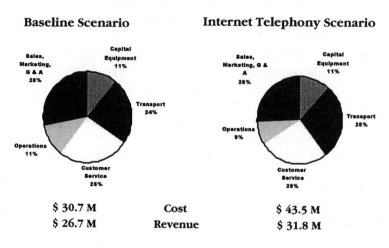

	Cost	
$ 30.7 M		$ 43.5 M
$ 26.7 M	Revenue	$ 31.8 M

Subscriber costs are affected in different ways. Table 10.3 shows the percentage increase in each cost category for each subscriber type. For example, transport costs increased by 75% for the analog dial-in subscribers. In general, transport costs are substantially affected for each subscriber type. Costs in the other expenses category increase for the leased-line subscribers (56kb and T1) due to an increase in sales and marketing costs. This is based on the assumption that leased-line subscribers would purchase enough capacity to maintain their circuit at the same level for both scenarios. Hence, additional revenue is received from the leased-line subscribers in the IT scenario. Because sales and marketing costs are based on a percentage of revenue, these costs also increase.

TABLE 10.3
Subscriber Cost Increase (Leida, 1998)

	Residential	Business	ISDN	56 kb	T1
Capital Equipment	45%	45%	80%	66%	63%
Transport	75%	75%	85%	64%	64%
Customer Service	44%	44%	44%	43%	44%
Operations	7%	7%	30%	26%	25%
Other Expenses	7%	7%	7%	78%	78%
Total	33%	34%	48%	59%	64%
Cost	$30	$32	$126	$745	$2,375

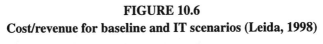

FIGURE 10.6

Cost/revenue for baseline and IT scenarios (Leida, 1998)

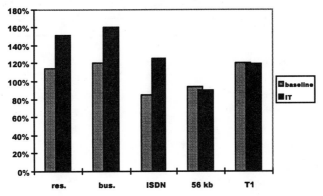

Drawing conclusions based on comparing the change in costs between the different types of subscribers is not valid because the revenue is also changing, but only for the leased-line subscribers. One method of comparing the impact on the different types of subscribers is to consider a cost/revenue ratio for both scenarios, which is presented in Figure 10.6. Here one can see that the dial-in subscribers become particularly unprofitable, but the leased-line subscribers remain at about the same level of profitability as in the baseline scenario.

Table 10.4 shows a breakdown of the 10 highest cost elements for residential dial-in subscribers. In the IT scenario, the cost elements for the dial-in subscribers maintain a similar distribution as for the baseline scenario. Two transport cost elements (analog dial-in T1 and T1 transport) increased, but only moderately. Nontechnical costs, such as customer service, sales/marketing, and general/administrative, together represent 59% of the per-user costs. Hence, as in the baseline scenario, nontechnical costs still play an important factor even in the face of IT.

As in the baseline scenario, (backbone) transport costs are still the major cost element for the T1 subscribers in the IT scenario (Table 10.5). Hence, those ISPs targeting T1 subscribers can gain substantial cost savings by having the most efficient network. This is in contrast to ISPs that target analog dial-in subscribers, for example, who would only receive marginal benefit by optimizing their network. For them, substantial cost savings can also be gained by targeting customer service and sales and marketing areas.

TABLE 10.4
High Cost Components for
Residential Dial-in Subscribers (IT Scenario) (Leida, 1998)

Item	Cost	% (IT)	% (Baseline)
Customer service	$10.80	36%	33%
Sales and marketing	$4.00	13%	18%
Analog modems	$3.51	12%	11%
General and administrative	$3.01	10%	11%
Analog dial-in T1	$2.68	9%	8%
T1 transport	$2.07	7%	4%
Billing	$1.25	4%	5%
POP personnel	$0.92	3%	4%
Software and hardware maintenance	$0.78	3%	2%
T3/OC-3 transport	$0.42	1%	4%
Total of top 10 items	$29.43	97%	—

TABLE 10.5
High Cost Components for
Residential T1 Subscribers (IT Scenario) (Leida, 1998)

Item	Cost	% (IT)	% (Baseline)
T3/OC-3 transport	$1,103	46%	46%
Sales and marketing	$398	17%	17%
General and administrative	$300	13%	13%
Customer service	$173	7%	8%
NAP Interconnection cost	$116	5%	5%
Cisco 7513 serial port card	$78	3%	3%
T1 CSU/DSU	$61	3%	3%
POP personnel	$44	2%	3%
Software and hardware maintenance	$37	2%	2%
Cisco 7513 router chassis	$33	1%	1%
Total of Top 10 Items	$2,343	99%	—

Sensitivity analysis of key parameters can provide further insight into the impact of IT on users' costs. For dial-in subscribers, modem costs are principally affected by usage patterns. In the IT scenario, it was assumed that call arrival rates and holding times each increased by 20% vis-à-vis the baseline scenario. Additionally, it was assumed that IT is used 33% of the time that a user is online.

These two parameters could be varied independently according to various scenarios.[16] Figure 10.7 shows how a dial-in user's total cost would change under these types of scenarios. The end result is that a dial-in user can impact his or her costs more by increasing total time online than by spending the same amount of time online while solely using IT. Similar analysis for customer service costs of dial-in subscribers shows that a doubling of customer service requests over the baseline scenario increases total subscriber costs by approximately 35%. Hence, any activity that prompts a user to spend more time online and to make more customer service requests will have a substantial impact on an ISP's costs. For T1 subscribers total costs are sensitive to changes in the usage patterns, which would seem intuitively correct because transport costs represent a large portion of T1 subscribers' costs.[17]

FIGURE 10.7
Internet telephony and usage level impact on subscriber's costs

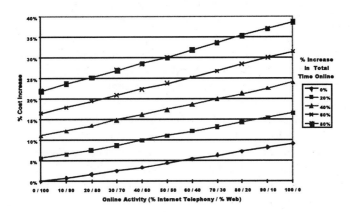

The IT scenario analysis yields the following conclusions:

1. In the IT scenario, the increase in the ISP's costs is double the revenue increase. Hence, ISPs will lose even more money if they do not attempt to recover additional costs.
2. Transport costs become the largest cost category in the IT scenario.
3. Nontechnical costs still remain a large portion of per-user costs, especially for dial-in subscribers.

---

16. The cost increase is relative to that of the baseline scenario. Because all other costs are held constant (such as customer service, sales, and marketing), the cost increase shown in this figure will not be the same as the for the IT scenario where other costs were assumed to increase.

17. See Leida (1998) for further sensitivity analysis.

4. Dial-in subscribers' costs are sensitive to access patterns and customer service costs. T1 subscribers' costs are sensitive to bandwidth usage patterns.

IT PRICING

As described previously, ISPs face potential increased cost pressure due to IT. To a lesser extent, they also face cost pressure due to access charge reform, which is discussed in the next section. For ISPs to remain in business, they will need to recover these increased costs. The Resource ReSerVation Protocol (RSVP) developed by the Internet Engineering Task Force (IETF) could be used as a mechanism for implementation of usage-sensitive pricing to recover those costs.[18] However, whether and where the current specification of RSVP is useful has not yet been determined. It is clear that RSVP by itself is not capable of resolving the host of network architecture and quality of service constraints on Internet pricing models. For example, how RSVP traffic could cross multiple networks has not been resolved.

When developing pricing schemes, service providers will have to look beyond the IT service and consider how to price differentiated and/or integrated Internet services generally. We argue elsewhere that an integrated regulatory framework will be required to permit the provision of such integrated services (Neuman, McKnight, & Solomon, 1997).

Alternatives for pricing include flat rate, or the introduction of usage-sensitive pricing. In McKnight and Bailey (1997), a variety of proposed approaches to Internet pricing, including approaches at the infrastructure level for network interconnection, are presented (see Bailey & McKnight, 1997). In the next few years, we anticipate experimentation with a variety of pricing models that permit service guarantees for multiple qualities of service, including guarantees for both real-time multimedia and multicast conferencing.[19]

Employment of yield management techniques, which may enable use of innovative service definitions in the face of highly variable demand to maximize revenue, should also be considered.[20] Yield management, which originated in the

---

18. IETF rfc 2205 (http://reference.nrcs.usda.gov/ietf/rfc2300/rfc2205.txt).

19. The announcement by America Online (AOL) in January 1998 that it was raising the price for its service to $21.95 per month from $19.95 per month suggests that at least one major ISP recognizes that the revenue–cost equation must be brought into balance, as we argue here.

20. For example, see Paschalidis, Kavassalis, and Tsitsiklis (1997). Ideally, a company using yield management wants to maximize its profit, not just its revenue. However, in most cases where yield management is currently used, the marginal cost of providing service is very small vis-à-vis fixed costs. Hence, maximizing revenue is essentially the same as maximizing profit.

airline industry and is discussed further in Leida (1998), uses a combination of service definition, pricing, and admission control. The fundamental principle of yield management is that different classes of service, be it Internet access or IT, are defined and only the high-priority classes are served during peak periods of demand. During low periods of demand, discount classes are intended to attract an increased level of demand. The consequence of such techniques is that the system's capacity is more full on average and revenues are higher.

Additionally, one must consider the state of technology when considering cost-recovery alternatives. Usage-sensitive pricing will not be an option until protocols that monitor the use of IT are deployed widely.

IT POLICY

There are a variety of policy issues raised by IT, none of which were addressed by the Telecommunications Act of 1996, which is a recipe for gridlock from a decade of litigation around the issues of the redefinition of market structures (Neuman et al., 1997). There appears to be a self-correcting quality to the degree to which policy frameworks can become misaligned with technical and economic conditions, but the time lag and social welfare loss may be substantial. Here, we focus particularly on a quantitative estimate of the costs for IT service providers. Additionally, we touch briefly on the regulatory discussions surrounding IT within the European Union.

The regulatory treatment of the Internet and IT service providers, in particular, has attracted substantial attention but little insight as yet. Broader discussion of a model for a new, convergent regulatory framework, with no distinctions between wireline and wireless, narrowband and broadband, broadcast and switched service, and content and conduit, may be found in Neuman et al. (1997).

Access Charge Reform

Telephony has been traditionally one of the most regulated industry segments in the United States. Under Federal Communications Commission (FCC) rules (specifically, the Computer II Inquiry) ISPs, being classified as 'enhanced service providers,' are exempt from regulations imposed on 'carriers,' such as long distance telephone companies.[21] These carriers must pay per-minute 'access charges' on the order of $0.06 per minute to the local phone companies who terminate each end of a long distance call (Werbach 1997). A trade association of telephone com-

---

21. In its "Computer II Inquiry" (FCC 1980) the FCC established the definition of (basic and) enhanced service providers and chose not to regulate them for reasons of public interest. Based on this definition, ISPs have always been classified as enhanced service providers (Werbach, 1997, p. 32).

panies—America's Carriers Telecommunications Association (ACTA)—filed a petition with the FCC asking it to regulate IT.[22] ACTA argued that ISPs providing IT services should pay access charges to the local telephone companies, as do other long distance service providers.

In May 1997 the FCC unveiled a reformed access-charge system.[23] Although not ruling explicitly on the ACTA petition, the FCC chose to not require ISPs to pay per-minute access charges.[24] Instead, the FCC imposed increased phone charges on business users—ISPs included—and residential users with a second phone line in the form of an increased subscriber line charge (SLC) and a new pre-subscribed interexchange carrier charge (PICC).

Under the new rules ISPs will see an increase in cost of their analog dial-in lines. The SLC went from a cap of $5.60 per line, per month to $9.00 on Januray 1, 1998 (although few carriers will be able to charge as high as the cap, the average has been calculated to be $7.61; FCC, 1997c), and the PICC goes from $0.53 to $2.75 per line, per month. Using the average charges, the impact on ISPs (or any multi-line business) will be a $4.23 per month increase for each analog line.[25] Plugging these updated costs into the ISP cost model yields an increase for the analog dial-in subscribers' cost for both the baseline and IT scenarios. Table 10.6 shows the initial results for the two scenarios compared to the results for the two scenarios with the access reform.

TABLE 10.6
Analog Dial-in Subscriber Costs for Four Scenarios (Leida, 1998)

	Baseline	Baseline with Access Reform	IT	IT with Access Reform
Capital equipment	$2.70	$2.70	$3.90	$3.90
Transport	$2.98	$3.44	$5.21	$5.86
Customer service	$7.50	$7.50	$10.80	$10.80
Operations	$3.07	$3.07	$3.27	$3.27
Other expenses	$6.52	$6.57	$7.01	$7.06
Total	$22.77	$23.27	$30.19	$30.89

An alternative method of access reform could have been to implement per-minute access charges for ISPs, as proposed in the ACTA petition. The effect of

22. See ACTA 1996.

23. See FCC 1997a and 1997b.

24. More generally, the FCC is trying to move away from the per-minute charges that were developed in the 1980s when telephone service was generally a monopoly to a system of flat-rate charges that will be more compatible with a competitive market.

25. ($7.61 − $5.60) + ($2.75 − $0.53) = $4.23.

such reform is shown in Figure 10.8. This analysis is based on the residential dial-in subscriber of the baseline scenario who spends 1,233 minutes per month online (approximately 41 minutes per day). The dial-in subscriber monthly cost is displayed as the per-minute access charge is varied. The result is that access charges quickly become the dominating cost element for a dial-in subscriber. ISPs would surely have to pass this cost increase on to the end user, which would have the effect of greatly impeding the continued growth in dial-in Internet services.

FIGURE 10.8
Access charge impact on dial-in user cost (Leida, 1998)

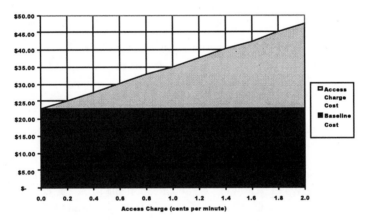

Although no cost increase is advantageous to ISPs, the recent FCC actions should be considered much less threatening than the potential impact of IT or of per-minute access charges. A principal conclusion that one reaches based on these cost results is that ISPs need either to prevent widespread use of IT, or to change the current pricing structure of Internet access services in order to recover the increased costs.

European IT Policymaking

As we have noted earlier, perhaps the greatest challenge for IT is how it will be treated by governments. The Internet is indeed growing in importance in the United States, and has therefore been focused on at the highest levels of the U.S. government more than in most other nations. However, it would be a mistake to ignore regulatory dilemmas and proposed approaches arising elsewhere. In particular, the European Commission's approach to determining policy for IT merits attention because of the obvious impact such policies may have in enabling, or inhibiting, the continued growth of a worldwide market for advanced Internet services.

Heterogeneity has been a key characteristic of the Internet from its beginning. The question of how much heterogeneity in Internet policy is tolerable for various classes of service will soon be answered in practice by policymakers and Internet users.

As discussed in Short (1997), the European Commission has established several criteria that IT must meet before it will be subject to regulation. These criteria, initially published in European Commission Directorate (1997), are presented in Table 10.7.

TABLE 10.7
EU Criteria for Voice Telephone (Short, 1997)

Such communications are subject of a commercial offer.
Such communications are provided for the public.
Such communications are to and from the public switched network termination points on a fixed telephony network.
Such communications involve direct transport and switching of speech in real time.

Based on these criteria, IT is not considered voice telephony because IT does not meet the criterion of 'real-time" communication due to the current, high levels of delay experienced by IT users on the public Internet.[26] Hence, IT services in Europe are not subject to regulation at this time.

CONCLUSIONS

IT service providers confront a variety of challenges. The costs, the technologies, business and pricing models, and the policy environment for IT are all unsettled and in a state of rapid evolution.

So far, a relatively hands-off policy approach has been taken by the FCC in the United States. A similar policy position has also been taken by the European Union. In spite of misguided efforts in some countries to ban IT, we believe the real challenge is how to align the costs, technologies, prices, and policies to enable a rich new class of differentiated and integrated Internet services to flourish, subsequently bringing substantial benefits to consumers. IT is acting as a catalyst to restructuring in the telecommunication industry. The rapid growth of new Internet Protocol (IP) based infrastructures, services, and applications resulting from these trends should benefit both consumers and producers worldwide, hastening the

---

26. Short (1997).

'creative destruction' of outmoded regulatory regimes, industry structures, and business practices.

In this chapter, we have described a cost model, developed by the ITC, of IT service providers. The model puts the ISP's costs into five categories: capital equipment, transport, customer service, operations, and other expenses (sales/marketing and general/administrative).

The model was evaluated with two scenarios: The baseline scenario, which represents an ISP today where its subscribers use primarily the Web; and the IT scenario in which the ISP sees a significant rise in the use of computer-to-computer IT. This scenario is in contrast to another potential IT scenario where a telephone service provider desires to replicate the PSTN using IT technology.

It was shown that with a moderate use of IT the increase in total ISP costs is nearly double the increase in revenues. Hence, ISPs, many of which are currently operating at unprofitable levels, would lose even more money if they fail to adopt new business models and change pricing policies to recover additional costs. Alternatives for cost recovery include various pricing and yield management techniques, some of which we have explored elsewhere (Leida, 1998; McKnight & Bailey, 1997).

The cost model presented and analyzed here provides a snapshot in time; conclusions (and the models on which conclusions may be drawn) must be reassessed, in real time, as technologies, industries, and regulatory environments evolve.

REFERENCES

America's Carriers Telecommunication Association (ACTA). (1996, March). *The Provision of Interstate and International Interexchange Telecommunications Service via the 'Internet' by Non-Tariffed, Uncertified Entities*, Petition to the FCC.

Bailey, J. P., & McKnight, L. W. (1997). Scalable Internet interconnection agreements and integrated services. In B. Kahin & J. H. Keller (Eds.), *Coordinating the Internet* (p. 309). Cambridge, MA: MIT Press.

Boardwatch Magazine Directory of Internet Service Providers. (1997, Winter).

European Commission Directorate-General for Competition (DGIV). (1997, July 5). *Notice by the Commission concerning the status of voice on the Internet under Directive 90/388/EEC*, OJ C 140 (p. 8). Available http://www.europa.eu.int/en/comm/dg04/lawliber/en/97c140.htm.

Federal Communications Commission. (1980). Amendment of §64.702 of the Commission's rules and regulations, final decision, 77 FCC 2d 384 (Second computer inquiry).

Federal Communications Commission. (1997a, May 8). *Federal-State Board on Universal Service*, CC Docket No. 96-45, First Report and Order, FCC 97-157 (Universal Service Order).

Federal Communications Commission. (1997b, May 7). *Commission reforms interstate access charge system*, Report No. CC 97-23.

Federal Communications Commission. (1997c, May 16). *Price cap performance review for local exchange carriers and access charge reform*, CC Docket Nos. 94-1 and 96-262, Fourth Report and Order in CC Docket No. 94-1 and Second Report and Order in CC Docket No. 96-262, FCC 97-159 (Access Reform Second Report and Order).

FIND/SVP. (1996, October). *Home Use of the Internet and Commercial Online Services*. Available http://etrg.findsvp.com/financial/homeuse.html.

Forrester Research, Inc. Telecom Strategies Group. (1996, August). *Sizing Internet Services*. Forrester Research: Cambridge, MA.

Leida, B. (1998). *A cost model of internet service providers: Implications for internet telephony and yield management*. Unpublished master's thesis, MIT, Cambridge, MA.

McKnight, L.W., & Bailey, J. P. (Eds.). (1997). *Internet economics*. Cambridge, MA: MIT Press.

Morgan Stanley (Technology/New Media). (1996, February). *The Internet Report*. Available http://www.ms.com.

Neuman, W. R., McKnight, L., & Solomon, R. J. (1997). *The Gordian knot: Political gridlock on the information highway*. Cambridge, MA: MIT Press.

Paschalidis, I., Kavassalis, P. & Tsitsiklis, J. N. (1997). Efficient resource allocation and yield management in internet services. In *Internet Telephony Consortium year-end report*. MIT: Cambridge, MA.

Short, K. I. (1997). *Towards integrated intranet services: Modeling the costs of corporate IP telephony*. Unpublished master's thesis, MIT, Cambridge, MA.

Students of the MIT Telecommunications Modeling and Policy Analysis Seminar. (1996). *A cost model of Internet telephony for regulatory decision making, in the matter of "the provision of interstate and international interexchange telecommunications service via the 'Internet' by non-tariffed, uncertified entities.'* TPP91, MIT. Available http://itel.mit.edu/docs/ACTA/TPP91.htm.

Werbach, K. (1997). *Digital tornado: The Internet and telecommunications policy* (OPP Working Paper No.29). Washington, DC: Federal Communications Commission Office of Plans and Policy.

Muddy Rules for Cyberspace

Dan L. Burk
Seton Hall University

INTRODUCTION

Several recent commentators on intellectual property generally,[1] and on digital media in particular,[2] have argued that adoption of "strong" or "complete" rights are necessary to properly protect such works, and that digital works in particular must be protected by such rights. Policymakers have adopted much the same theme.[3] These claims have profound implications for development of markets in digital works. In the print world, property rights in creative works have typically been incomplete, and subject to public interest takings, such as fair use. The call for complete rights represents a radical departure from established practice, and is particularly troubling in a networked environment such as the Internet, where complete rights would allow control over every possible usage.

The analyses on which these proposals are based draw heavily on analogies to rights in real property, especially the economic analysis of rights in real property. Yet, curiously, none of these "strong property" analyses deal seriously with studies addressing the legitimate role of other kinds of entitlements in real property. In particular, there is little or no discussion of the literature addressing the role of unclear or "muddy" entitlements.[4] This essay suggests to the contrary a role for

---

1. Kenneth W. Dam, *Some Economic Considerations in the Intellectual Property Protection of Software*, 24 J. LEGAL STUD. 321 (1995); Frank H. Easterbrook, *Intellectual Property Is Still Property*, 13 HARV. J.L. & PUB. POL'Y (1990).

2. *See* I. Trotter Hardy, *Property (and Copyright) in Cyberspace*, 1996 U. CHI. LEGAL. F. 217.

3. *See, e.g.*, INFORMATION INFRASTRUCTURE TASK FORCE, INTELLECTUAL PROPERTY AND THE NATIONAL INFORMATION INFRASTRUCTURE: THE REPORT OF THE WORKING GROUP ON INTELLECTUAL PROPERTY RIGHTS (1995).

"muddy" rules in cyberspace. It first reviews the case for complete property rights in informational works, highlighting certain difficulties with such arguments. It then offers a set of critiques that suggest an important role for muddy entitlements in the law of intellectual property, and which challenge clear entitlements as necessarily the optimal rule for intellectual property. The chapter then turns to a discussion of the Internet, showing that its idiosyncracies render the complete property argument even more suspect in cyberspace than it is in real space. Finally, it suggests certain reasons that "muddy" entitlements may be beneficial to fostering informational works in the online world.

DEFINING PROPERTY

Recent intellectual property scholarship draws heavily on the nomenclature of "property rules" and "liability rules" articulated by Calabresi and Melamed in their famous framework of rights and responsibilities.[5] The two types of rules are primarily differentiated by the ability of the property owner, under a property rule, to exclude others from use of the property. Under a liability rule, the owner cannot exclude others from taking or using the property, but he or she can demand compensation or damages. Decisional authority distinguishes the two systems: Under property rules, the owner makes the decision to exclude or not; under liability rules, the option to take or not rests with outside parties.

This distinction between property rules and liability rules is not precisely the same as the distinction between complete and divided entitlements,[6] although the two may overlap.[7] Divided entitlements appear whenever more than one entity has a claim to the property; that is, where the property owner must share or cede some uses of the property under some circumstances. This divided entitlement might be subject to either a property rule or a liability rule; it may even oscillate between the two—for example, under normal circumstances, a boat owner must have permission to tie up to a dock, but in a storm, he or she may be permitted to tie up without permission, so long as he or she pays for any damage caused. The essential question

---

4. *See, e.g.*, Carol M. Rose, *Crystals and Mud in Property Law*, 40 STAN. L. REV. 577 (1988); Thomas Merrill, *Trespass, Nuisance, and the Costs of Determining Property Rights*, 14 J. LEGAL STUD. 13 (1985).

5. Guido Calabresi & A. Douglas Melamed, *Property Rules, Liability Rules, and Inalienability: One View of the Cathedral*, 85 HARV. L. REV. 1089 (1972).

6. A. Mitchell Polinsky, *Resolving Nuisance Disputes: The Simple Economics of Injunctive and Damage Remedies*, 32 STAN. L. REV. 1075, 1087 (1980).

7. Ian Ayres & Eric Talley, *Solomonic Bargaining: Dividing a Legal Entitlement to Facilitate Coasean Trade*, 104 YALE L. J. 1027 (1995); Louis Kaplow & Steven Shavell, *Property Rules Versus Liability Rules: An Economic Analysis*, 109 HARV. L. REV. 713 (1996).

for a divided entitlement is not so much *if* the owner can exclude or merely demand a royalty, but rather *when* can the owner exclude or demand a royalty.

Finally, entitlements may be subject to muddy or to clear legal rules.[8] This third distinction has reference to the rules demarcating the extent of an entitlement; the demarcation is said to be clear if the outcome of an entitlement dispute is predictable. If it is not, the rule is said to be muddy. Muddy entitlements tend to be subject to more complex legal tests, that are said to give the decision maker flexibility in determining the outcome. This is not to say that a complex rule is necessarily a muddy rule.[9] Rather, the hallmark of a muddy entitlement is its subjectivity—typically, it involves some type of legal balancing test, whereby a court or decision maker weighs a variety of competing factors before coming to a determination of entitlements in the property. However, the entitlements themselves could be either Calabresi–Melamed property rights or liability rights, and could just as well be either divided or complete.

These different types of entitlement rules may overlap, in that they sometimes achieve similar effects. For example, a liability rule might under certain circumstances be considered a type of divided property right, as an outside party can exercise an "option" to use the property, subject to a fee.[10] Additionally, a divided entitlement might well be created or approximated by a muddy standard—because the entitlement is muddy, the parties to a disputed property right will be uncertain as to the extent of their rights. This causes them to view the entitlement in a probablilistic manner, which itself constitutes a type of divided property entitlement.

Justifying Complete Rights

Some recent commentary on intellectual property adopts the position that intellectual property entitlements must be as complete as possible. The "completeness" of a bundle of rights would be maximized by adopting bright-line property rules that appropriate the full value of the holding to the owner. What rationale might justify adoption of such rules? Intellectual property rights are most often justified as solutions to the public goods problem: Informational works resemble public goods in being nonrival, and the easy reproducibility of such works similarly mimics the nonexclusivity of public goods.[11] Consequently, because the marginal cost of reproducing and distributing such works is extremely low, there will be little incentive to create such works if they are sold at a price near the marginal cost.[12] Legal

---

8. *See* Rose, *supra* note 4 (describing clear and muddy entitlements).

9. *See* Merrill, *supra* note 4 at 23 n. 43.

10. *See* Ayres & Talley, *supra* note 7 at 1041.

11. Paul A. Samuelson, *The Pure Theory of Public Expenditures*, 36 REV. ECON. & STAT. 387 (1954).

12. *See* William M. Landes & Richard Posner, *An Economic Analysis of Copyright Law*, 18 J. LEGAL STUD. 325, 326 (1989).

fences, in the form of exclusive rights, allow the creators of such works to exclude free-riding patrons of their work and to extract a payment for access to the work. The ability to extract payment offers an incentive that encourages the creation and dissemination of such works.[13]

This argument offers a plausible rationale for the creation of intellectual property rights. Yet it offers poor grounds for complete intellectual property rights, because the rationale implies an efficiency trade-off. As the recipient of the intellectual property right raises the price of the work beyond the marginal cost near zero, people who would have chosen to purchase or obtain the work at the lower price can no longer afford it. Thus, the incentive to create the work is purchased at the expense of restricted availability.[14] That restricted availability represents a loss of social welfare. The loss is acceptable up to the point required to induce creation of the work, but not beyond. A property right that appropriates all the value of the work to the creator is not always necessary to induce creation of the work; presumably, the creator would be prompted to create if he or she received a right that ensured he or she could at least cover fixed costs.

Consequently, complete property rights require additional justification when the welfare loss from increasing the cost of the work exceeds that necessary to prompt its creation. One alternative is the *revealed preference* justification. Under this theory, intellectual property rights are necessary to determine what level of investment and production in such works is desirable.[15] Buyers' willingness to pay acts as a signal to producers, telling them how much of the good society desires, and hence what level of resources should be devoted to producing the good. The property owner should have complete rights to the work in order to receive complete information about its demand.

A major difficulty of this rationale is that the signal as to the correct level of production is skewed from the outset by the nature of the goods in question. For tangible goods, pricing at marginal cost places the price of the goods within reach of the largest feasible number of buyers and sets the correct level of output.[16] Not so for intellectual goods, where pricing at marginal cost means pricing at zero or close to zero. Legal exclusion allows the producer to price above marginal cost, but pricing above marginal cost excludes a number—perhaps a large number—of buyers who would otherwise purchase or acquire the good. Consequently, the true dimensions of the potential market for the good are obscured, and we cannot get a correct picture of consumer preference for the good after all.[17]

---

13. *Id.*

14. ARMEN ALCHIAN & WILLIAM R. ALLEN, EXCHANGE & PRODUCTION: COMPETITION, COORDINATION & CONTROL 100 (3d ed., 1983).

15. *Id.*

16. *Id.* at 208–09.

17. *See* Edmund W. Kitch, *Patents: Monopolies or Property Rights?*, 8 RES. L. & ECON. 31 (1986).

A third rationale might be termed the coordination rationale. This theory holds that complete control over the information good must be given to the producer in order for him or her to manage efficient use of the resource.[18] If control of the good is incomplete or held in common, conflicting uses might arise. Of course, nonrival goods cannot be congestible in the same way that rival goods are, so the "conflicting uses" of informational goods must be somehow different than for physical goods. The supposedly conflicting uses are the unwillingness of buyers to invest in a good that might also be developed by someone else—for example, a producer might be reluctant to invest in a feature film based upon a novel if another producer could be doing the same. By giving the novelist full control over the work, he or she is able to raise the price of the good, and locate the buyer who values it the most—that is, the buyer who is willing to pay the most for the exclusive right to develop the informational good.[19]

The trouble with this argument is that it brings us full circle—it is simply the first rationale once removed, where an exclusive right is needed to induce the buyer to buy, rather than to induce the producer to produce. Once again, a welfare trade-off occurs between the exclusivity necessary to induce the highest value developer to develop the work, and the other developers who in theory could simultaneously hold and use the nonrival work.

MUDDY RULES FOR COPYRIGHT

These rationales for complete entitlements in copyright, and particularly the coordination rationale, are based on analogies to the law of real property. Yet the law of real property is not at present, nor has it ever been, subject to an exclusive regime of complete, bright-line Calabresi–Melamed property entitlements. Instead, real property encompasses a variety of entitlement regimes. It may be, of course, that the real property system is flawed, and should also be entirely converted to complete entitlements. However, the variety of rules in real property has been analyzed to consider when differing rules may be appropriate to address different transactional environments.

Examples of Muddy Rules

The classic real property example of muddy rules occurs in the law of nuisance. The boundaries of land may appear to be clearly demarcated, and the owner's right to exclude others from its use absolute. Yet, if enjoyment of the property is curtailed by a neighbor's loud noises, drifting smoke, noxious odors, or shining light, the

---

18. *See* Edmund W. Kitch, *The Nature and Function of the Patent System*, 20 J.L. & ECON. 265 (1977).

19. *Id.* at 276.

property owner has no right to prevent the invasion unless he or she can prove that the neighbor's activity is "unreasonable"under the law of nuisance—that the value of the disturbing activity is less than the harm it is causing.[20] Thus, the right to exclude such invasions is unclear: Depending on the social utility of their respective activity, either the property owner or the neighbor may hold the entitlement.

The emergence of unclear nuisance laws in real property is believed to be a response to situations in which costs for transacting bargains are prohibitively high.[21] Where transaction costs are low, real property rules such as trespass give the property holder a clear right to exclude, which facilitates bargaining for the right. However, where transaction costs are high, such bargaining will not occur in any event, and the clear right simply reinforces the impasse: Property remains with the initial rights holder even though others place a higher value on it. Development of an unclear or muddy rule helps break the impasse by encouraging either less costly informal bargains or by moving the dispute to court for a third-party review of the entitlement. Because the ownership of the right is unclear, claimants are forced to deal with one another in one fashion or the other.

Copyright doctrines such as fair use appear to operate similarly to nuisance.[22] Fair use, in particular, has been identified as an area in which the transaction costs associated with negotiating the taking of a copyrighted work would tend to either exceed the value of the taking or to frustrate takings of high social value.[23] Fair use is therefore subject to a legal balancing test to determine whether the unauthorized use of the work was socially desirable. Because the allocation of entitlements under fair use is ambiguous, the standard will sometimes channel creators and users into court to determine ownership. The muddy four-part balancing standard of fair use allows courts to reallocate what the market cannot.[24] More often, the uncertainty created by the muddy standard may tend to channel buyers and sellers into less costly informal structures. In the case of fair use, this may primarily take the form of a truce between owners and users when the taking stays below a certain threshold. In cases where the taking is more substantial, repeated, or valuable, litigation will occur to produce a custom-made allocation for each situation.

Much the same might be said for originality, a requirement for copyright authorship. The test for originality is decidedly ambiguous; courts have developed exceptionally murky tests of "abstraction" and "filtration" to separate out elements that are and are not copyrightable, and then compare the copyrightable elements to see if they are "substantially similar." Almost no one, including experienced

---

20. *See, e.g.,* RESTATEMENT (SECOND) OF TORTS § 826 (describing balancing test for legal nuisance).

21. *See* Miller, *supra* note 4.

22. RICHARD A. POSNER, ECONOMIC ANALYSIS OF LAW 53 n. 3 (4th ed. 1992).

23. *See* Wendy Gordon, *Fair Use as Market Failure: A Structural and Economic Analysis of the Betamax Case and its Predecessors,* 82 COLUM. L. REV. 1600 (1982).

24. POSNER, *supra* note 22.

copyright attorneys, can determine in advance what the outcome of such a test might be, because it is highly case specific. This is another classic example of a muddy entitlement, and it appears to have evolved in this fashion for much the same purpose as fair use. In many instances, it will encourage informal allocations among creators and subsequent users, but at some threshold, litigation to allocate ownership will occur.

Shared Property Rights

As noted earlier, muddy entitlements may function in much the same way as a divided entitlement. Consequently, a prescription for muddy entitlements may resemble the various "shared property" analyses that have appeared in the recent literature on property and intellectual property. From these analyses, we may glean some additional reasons as to why muddy entitlements exist in copyright. They suggest that muddy rules are beneficial because they dampen strategic behavior, increase the gains from bargaining, and encourage positive externalities.

One noted analysis of divided rights is the Solomonic Bargaining argument proposed by Ayres and Talley. Ayres and Talley argued that divided entitlements will prompt market participants to forgo some types of strategic bargaining. This occurs because of each party's rational self-interest: Given that the entitlement is divided, either party may ultimately prove to be the purchaser rather than the seller of the property, depending on the path negotiations take. Thus, neither party would wish to understate their valuation of the property, as it may prompt an offer to buy at that price.[25] Ayres and Talley went on to argue that divided or "fractional" entitlements will also facilitate trade by reducing parties' incentives to bargain strategically.[26] Although they did not explicitly discuss muddy rules, they pointed specifically to "probabilistic" entitlements that may dampen strategic behavior.[27] Such probabilistic entitlements may, as indicated earlier, include muddy entitlements, suggesting that muddy rules may tend to facilitate bargains by curbing strategic behavior.

Another recent look at divided entitlements was offered by Lemley in his discussion of "blocking" patents and of copyrights.[28] Lemley argued that divided entitlements, such as those found in patent law as "blocking" patents, serve to facilitate improvements in intellectual property. This is not necessarily a result of lessened strategic bargaining, as in the argument of Ayres and Talley. Rather, Lem-

---

25. *See* Ayres & Talley, *supra* note 7.

26. *See id.* at 1073–74.

27. *Cf.* Jason Scott Johnston, *Bargaining Under Rules Versus Standards*, 11 J.L. ECON. & ORG. 256 (1995) (discussing comparative benefits of bargaining over muddy entitlements).

28. Mark A. Lemley, *The Economics of Improvement in Intellectual Property*, 75 TEX. L. REV. 989 (1997).

ley suggested that divided entitlements increase the potential gains from trade:
When each party has a larger stake in the outcome of a negotiation, the negotiation
is more likely to occur. By contrast, if complete control is given to the original creator of a work—as Lemley argued is currently the case in copyright—there is little
for original developers and improvers to bargain over; the original developer holds
all the cards, and there is no incentive for an improver to do any improving.

Although Lemley's argument is offered largely in terms of divided entitlements,
a muddy entitlement rule may serve much the same purpose. This has some important implications both for his argument and for the analysis here. Copyright
may lack a parallel doctrine to that of blocking patents, but infringement doctrines
such as abstraction/filtration and substantial similarity, together with fair use, create uncertainty as to the extent of an author's entitlement. Such uncertainty makes
enforcement a doubtful proposition except perhaps in the most egregious cases of
slavish copying. Even if an enforcement suit is brought, the muddy doctrines leave
the court considerable latitude to excuse socially valuable copying. Consequently,
it may be that improvement in copyright is not as impeded as we might at first expect because such improvement is facilitated by the presence of muddy rules.

An additional recent analysis deserves attention, that of the "democratic paradigm" articulated by Netanel.[29] This is not a shared property analysis, at least not
in the sense of the Ayres–Talley and Lemley analyses. However, it provides a related rationale for embracing muddy entitlements in at least some instances. Netanel's discussion is not couched in economic terms, and one suspects that he might
object to its reformulation in such a fashion: The democratic paradigm, he informed us, is in the market but not of the market.[30] Nonetheless, it appears that the
gist of his argument can be encompassed as an economic argument. According to
Netanel, copyright ownership should be calibrated in such a way that it serves to
foster democratic ideals. In essence, Netanel appears to suggest that significant
positive externalities are generated by open access to informational works.

This view indicates an additional role for muddy intellectual property entitlements. Copyright holders are unlikely to consider such beneficial spillover effects
in choosing their price and level of production. Consequently, the statute is designed to allow considerable latitude to users or consumers in taking from copyrighted works, particularly where the social value of the taking is high and
transaction costs are prohibitive. The muddy copyright entitlements considered
here may facilitate such valuable takings. First, by obscuring the boundaries of the
copyright holders' entitlement, a muddy rule discourages attempts at enforcement
where the unauthorized taking is small; the expected return from litigation will be
too small to justify the expenditure in enforcement. Second, for larger or more ex-

---

29. Neil Weinstock Netanel, *Copyright and a Democratic Civil Society*, 106 YALE L.J.
283 (1996).
30. *Id.* at 386.

tensive takings, the muddy standard channels the dispute into court, where a third-party arbiter can take into account the public value of the taking in rendering a decision on infringement. Thus, muddy standards may facilitate bargaining where the community "shares" a portion of the entitlement as external benefits.

CYBERSPACE TRANSACTION COSTS

The previous section sketches out a series of reasons for which muddy entitlements have been and continue to be appropriate to govern the disposition of intellectual property in creative works. However, these arguments, even if accepted, do not necessarily prove the case for muddy rules in cyberspace. Some have argued that conditions in cyberspace may be more favorable to clear, undivided property entitlements than are conditions in "real space."[31] This would occur if cyberspace transaction costs are relatively low. However, on closer examination, the situation may prove to be quite the opposite: Not only will transaction costs in cyberspace be at least as high as those in real space, transaction costs online may well be considerably higher.

Determining Borders

The first argument for property entitlements in cyberspace rests on the assumption that property in cyberspace can be clearly demarcated.[32] In the case of real property, Demsetz[33] and Ellickson[34] have argued that property rights are most appropriate where clear borders can be easily and cheaply drawn—or, in other words, where borders are cheap, people will "buy" more of them.[35] In such an environment, recognizing, policing, and transferring property is simpler because clear demarcations are readily available. This theory of demarcation has clear implications for the choice between muddy standards and clear rules, as adoption of muddy standards is more likely to be indicated where the cost of demarcation is high.

Hardy argued that cyberspace offers conditions under which demarcation is cheap. He observed, for example, that information on the Internet can be easily bordered by the designation of a computer file—indeed, this is how computer information systems are now structured.[36] Thus, Hardy suggested that the boundaries of a digitized work, such as a software file, can be easily determined, and so digital media should be conducive to the development of strong property entitlements.

---

31. *See* Hardy, *supra* note 2.
32. *Id*. at 242.
33. Harold Demsetz, *Toward a Theory of Property Rights*, 57 AM. ECON. REV. 347 (1967).
34. Robert C. Ellickson, *Property in Land*, 102 YALE L.J. 1315 (1993).
35. Hardy, *supra* note 2 at 234.
36. *Id*. at 243–44.

However, the matter of determining boundaries may be somewhat trickier than simply looking for the last string of binary code in a certain file. Just as we can find the beginning and end of a computer file, so too we can find the boundaries of a piece of art or of a text. Works found in real space are bounded by clear borders: frames, margins, or bindings. However, simply determining such a boundary is not—and has never been—a sufficient criterion for designating the "boundaries" of a copyrightable work. These borders are relevant only to the delineation of the embodiment copy of the work, not to delineating the borders of the work itself. Indeed, this is precisely why we need legal rules to "fence" intellectual property in the first instance: because of the difficulty of discerning a natural or physical boundary to the property.

It admittedly might be simpler to merely determine the edges of a painting or a piece of text and declare that all the work—original or not—within the border was the exclusive intellectual property of the possessor of the painting. However, such a boundary is not relevant to the goal of fostering creative works. It would be absurdly simple to similarly deploy a bright-line ownership rule such as "Anything written on a Thursday can be reproduced subject to a compulsory license fee." Those boundaries are arbitrarily chosen, and simply do not reflect the goals for which we allow ownership in information.

Determination of the proper copyright boundary in any given instance is a costly and complex proposition and not to be undertaken lightly. A full discussion of the reasons that demarcation is costly would run far beyond the scope of this chapter, but one reason deserves particular mention. A growing body of copyright scholarship reminds us that authorship is a complex process, to which the reader may bring as much meaning and value as the originator.[37] Thus Coombe wrote of how readers "recode" text to yield a work that is the result of both reader's and writer's interpretation.[38] Aoki and Rotstein referred to such recoded meanings as the "intertext" of a work.[39] In the work's intertextual spaces, the respective contributions of author and reader are unclear and perhaps inseparable.

These observations on intertext and recoding may seem somewhat far removed from the practicalities of property ownership. Even if a reader does recode a work or add meaning to it, this occurs largely in the reader's mind, where the law does not inquire into the reader's unfixed contribution to the work. However, as Chon observed, where digital works are concerned, the process of recoding quickly

---

37. *See, e.g.,* JAMES BOYLE, SHAMANS, SOFTWARE, AND SPLEENS: LAW AND THE CONSTRUCTION OF THE INFORMATION SOCIETY (1996); Peter Jaszi, *Toward a Theory of Copyright: The Metamorphoses of "Authorship",* 1991 DUKE L.J. 455 (1991).

38. Rosmary Coombe, *Objects of Property and Subjects of Politics: Intellectual Property Laws and Democratic Dialogue,* 69 TEX. L. REV. 1853, 1863 (1991).

39. See Robert H. Rotstein, *Beyond Metaphor: Copyright Information and the Fiction of the Work,* 68 CHI-KENT L. REV. 725 (1993); Keith Aoki, *Adrift in the Intertext: Authorship and Audience Recoding Rights,* 68 CHI-KENT L. REV. 805, 810 (1993).

spills from carbon-based organic memory into silicon-based mechanical memory.[40] Chon cited the example of electronic mail "discussions" in which portions of previous messages are incorporated into replies.[41] The result is something different than a derivative work or a joint work.

Similarly, much of the copyright controversy surrounding inline framing of Web materials stems from the difficulty of neatly parsing the displayed result. Cavazos persuasively showed that under traditional copyright analysis, inlining results in no direct infringement, contributory infringement, or vicarious liability: The framed materials are called directly from the content provider's server, and never pass through the framer's server.[42] Yet the displayed work appears to be a different work from the one contemplated by the content provider. The "authorship" of the displayed result is unclear; under traditional analysis, it is perhaps authored by no one.

Such examples suggest that formulating any sensible demarcation of intellectual property in cyberspace may be a much more expensive undertaking than demarcation of intellectual property in physical space—and the latter is a fairly high-cost proposition to begin with. Although we could demarcate the ownership of cyberspace information with a clear and unambiguous rule—e.g., "every other line of code belongs to the reader"—that would hardly advance any of the plausible purposes for having intellectual property entitlements, and particularly not those professed by the advocates of complete entitlements.

Search and Bargaining Costs

An additional set of relevant transaction costs are those related to search and negotiation. In order for property to change hands, buyers and sellers must be able to locate one another and reach some accommodation. These costs of search and negotiation can be problematic in real space, but the current state of affairs on the Internet suggests that the costs of online transactions may be prohibitive. Simply finding online information that one might like to use or purchase is a major undertaking—a common metaphor for the Internet compares it to the contents of the Library of Congress without call numbers and dumped out on the floor. In such an environment, finding a piece of information worth purchasing can be a discouraging task. Subsequently, finding the owner from whom to make the purchase is equally daunting.

Technology, we are told, may help to alleviate some of the search costs associated with matching authors and users.[43] This might be accomplished using greatly

---

40. Margaret Chon, *New Wine Bursting From Old Bottles: Collaborative Internet Art, Joint Works, and Entrepreneurship,* 75 ORE. L. REV. 257, 271–72 (1996).

41. *Id.* at 261.

42. *See* Edward A. Cavazos, *Potential Linking Liability on the World Wide Web,* 42 BUFFALO L. REV (forthcoming 1997).

improved search engines, software agents, or other sophisticated searching tools. Of course, such tools are not currently available. It is for this reason that Internet users jealously gather and guard their precious cache of bookmarks—there is no other way to assure finding a resource again. Neither is it clear when such amazing search tools will become available, nor what they will cost when they do become available. The price for such tools will have to be exceptionally small if every chunk of information on the Web is to be tagged and retrievable.

Even assuming that in the not-too-distant future, every byte of information is tagged and monitored at a cost low enough to make tagging and monitoring worthwhile, such a system would only solve the problem of locating proprietary information and its identifying its owner. This would be only the first step in solving the problem of search costs. Under the most sophisticated rationale for complete entitlements, the purpose of creating the rights is to allow coordination of usage by a self-interested property owner. In order for this to happen, not only must potential buyers and sellers be able to locate one another, but the seller must be able to identify the buyer who places the highest value on the work.

To be sure, if a universe of potential buyers can be identified, then the property owner is a step closer to identifying the subset of buyers with the highest subjective valuations. However, the cost of selecting from among potential buyers may itself be prohibitive. Both buyers and sellers have strong incentives to cloak their true preferences in order to capture a larger share of the gains from trade. Such strategic behavior to either conceal or misrepresent one's true preferences impedes negotiations, raising the cost of reaching an agreement.

An early version of this problem was identified by Froomkin and DeLong in their survey of emerging electronic markets.[44] They employed a primitive software agent called Bargain Finder that can be programmed to query online vendors of music CDs in order to locate the best price for a given item. However, when they attempted to use the agent, Froomkin and DeLong found that many vendors refused access to their publicly available information on prices. Although we cannot be certain precisely why vendors might block the operation of such a "shop-bot," this conduct fits the expected outcome for a model of strategic behavior rather than that of pure price-based competition. The merchant may be reluctant to reveal a price for the item without being able to gauge the potential buyer's willingness to pay. Alternatively, merchants may be anxious to encourage repeat business through nonprice incentives, such as better service, which would not be communicated to the buyer via "shop-bot."

---

43. Hardy, *supra* note 2 at 237.

44. *See* J. Bradford DeLong & A. Michael Froomkin, *The Next Economy?* in INTERNET PUBLISHING AND BEYOND: THE ECONOMICS OF DIGITAL INFORMATION AND INTELLECTUAL PROPERTY (D. Hurley, B. Kahin & H. Varian eds. 1998).

When this problem is considered in the context of electronic commerce, one soon realizes that the Internet might almost have been designed to facilitate strategic behavior. The difficulty of locating an e-mail address is only the prelude to the kind of personal obscurity the Internet can bring. Online negotiations are stripped of much of the context that buyers and sellers might normally employ either to signal preferences or to infer preferences. Buyers and sellers may even choose to conduct negotiations through anonymous remailers or anonymous web sites. The now hackneyed cartoon from the *New Yorker* proclaims that "On the Internet, no one knows you're a dog"—but similarly, on the Internet, no one knows that you are a multimillionaire or a deadbeat.

Of course, what technology can cloak, technology can also reveal. Much of the current concern over electronic privacy stems from this concern: Merchants may gain the "upper hand" in the strategic bargaining game by extrapolating personal preferences from electronic records of consumers' online activity. Consumers might find offers tailored to their particular tastes and abilities to pay, but they might also lose consumer surplus to producers in the process. Equally important, they will lose strategic bargaining leverage. If strategic behavior by both parties to a transaction creates an impasse, creating a massive informational asymmetry in favor of one party will likely exacerbate the problem.

Jurisdictional Uncertainties

Because the Internet spans national borders, online activity creates a variety of jurisdictional and choice of law problems. This state of affairs presents formidable barriers to online negotiations over digitized materials. No clear rule of decision exists to determine the law that should be applied when information products are transmitted from one jurisdiction to another. Copyright can vary substantially among jurisdictions, and each jurisdiction's copyright regime will entail specific costs in terms of rights conferred or surrendered, procedural requirements, enforcement, and so on. Buyers and sellers of informational products will need to take such details into account in determining the total "cost" of any intellectual property bargain—purchase of copyrighted materials is in fact purchase of a bundle of rights associated with those materials. Without knowing the parameters of those rights, buyers and sellers simply will not know the value of what they are buying and selling.

Of course, buyers and sellers may attempt to contract with one another regarding choice of law and choice of forum. Two obvious choices for such contractual provisions might be the law of the jurisdiction of origin and the law of the jurisdiction of receipt. However, negotiating such provisions for every use of online resources is itself burdensome and a default rule may be desirable. In real space transactions, it would be highly unusual—and burdensome—to negotiate choice of forum and choice of law each time one made a purchase at the bookstore, the supermarket, the dry cleaner and so on. We habitually rely on default rules of geo-graphic jurisdiction that are not available in cyberspace.

Some commentators have suggested that adoption of a straightforward rule, such as "always use the law of the jurisdiction of receipt"or "always use the law of the jurisdiction of origin" would simplify matters.[45] In fact, clarity of such a rule is illusory, at least for purposes of negotiation: Buyers may not be able to determine in advance the location of the materials accessed, and so cannot de-termine the legal "cost" associated with the materials under a rule that adopts the law of the jurisdiction of origin. Similarly, sellers may not know the location of buyers, and so cannot calculate the correct price for the goods under a rule that adopts the law of the jurisdiction of receipt. Consequently, jurisdictional factors pose an additional cost to online transactions, either as negotiated items or as matters of uncertainty.

THE BENEFITS OF AMBIGUITY

If transaction costs in cyberspace remain as high as those in real space, and in some instances higher, then we must consider seriously retaining the types of muddy entitlements found in real space intellectual property regimes.

Informal Solutions

As indicated earlier, one purpose for adopting such muddy standards is to shunt buyers and sellers of property into informal negotiating systems, particularly where the transactional cost of formal negotiations will be high. The concept of such informal or self-help systems as an alternative to strong property rights has a long tradition in the discussion of copyright law. Palmer argued, for example, that supposedly public goods can frequently be financed without the creation of a state-sponsored exclusive right.[46] Palmer reminded us of Coase's discovery that the cost of lighthouses—supposedly a classic example of a public good—was in fact internalized by a clever system of private rents.[47] Initially, the direct benefits of a lighthouse would seem to be both nonrival and nonexcludable; any ship within sight of the lighthouse could use it as a warning beacon whether the ship pays for the light or not. In fact, however, fees for maintenance of lighthouses could be ex-tracted from ships docking in adjacent ports—indeed, such ships would have de-rived the greatest benefit from the beacon, and would likely value it the most

---

45. *See, e.g.*, Jane Ginsburg, *Putting Cars on the Information Superhighway: Authors, Exploiters, and Copyright in Cyberspace,* 95 COLUM. L. REV. 1466 (1995).

46. Tom Palmer, *Intellectual Property: A Non-Posnerian Law & Economics Approach,* 12 HAMLINE L.J. 261 (1981).

47. See Ronald Coase, *The Lighthouse in Economics,* 17 J.L. & ECON. 357 (1974).

highly. Consequently, by "tying" the use of the port to the maintenance of the lighthouse, a private source could be used to subsidize a public good.

Palmer suggested that similar alternatives may be available in the case of informational goods, obviating all or most of the need for copyrights. Examples of such arrangements abound in industries that nominally enjoy copyright protection, but in which such protection is very difficult to police: software publishers offering "upgrades" or customer service for software, or text publishers offering updates or "pocket parts" for printed treatises. Similarly, innovators in markets in which copying is inevitable and uncontrollable may "tie" easily copied works to goods or services that cannot be easily copied. This results in the so-called "Netscape" business strategy, named after the popular Web browser that employed it, in which distribution of the first product assists in seizure of "mindshare" but actual profits are made from sale of the second more controllable product.

These principles have been explicitly extended to cyberspace by commentators such as Barlow and Dyson.[48] In a set of widely cited and highly controversial articles, both Barlow and Dyson argued that the easy reproduction and distribution of digital works will disable copyright as we know it. In the absence of effective legal recourse, they contended, content producers will rely on other methods to extract value from their labors. Their list of nonlegal alternatives includes possibilities such as those suggested by Palmer.

This possibility of noncopyright alternatives to protect creative works has been compared by Hardy to a pie with several slices comprising the necessary protection.[49] He identified a copyright slice, and other slices as well—a contract slice, which offers an alternative form of legal protection; a "natural barriers" or state of the art slice, which comprises the difficulty of copying works in a particular medium; and a technological protection slice, which allows the erection of additional barriers to copying. Hardy argued that if any of these slices shrinks, others must expand to fill the empty spaces. He suggested, for example, that in the context of digital media, the "natural barriers" slice has shrunk, as copying bits is exceptionally easy. He therefore suggested that technological barriers, copyright, or contract be relied on to fill the resulting breach.

Hardy and Palmer are therefore in agreement that alternatives to copyright are both available and necessary. But the Palmer analysis, supplemented by Barlow and Dyson, suggests that Hardy's analysis is several slices short of a pie. We have already shown one slice that Hardy did not consider—the "tying slice"—but

---

48. *See* John Perry Barlow, *The Framework for Economy of Ideas: Rethinking Patents and Copyrights in the Digital Age* WIRED Mar. 1994 at 83 ; Esther Dyson, *Intellectual Value*, WIRED July 1995 at 136.

49. Hardy, *supra* note 2 at 223-34; I. Trotter Hardy, *Contracts, Copyrights, and Preemption in a Digital World*, 1 U. RICH. J.L. TECH, art. 2 (April 10, 1995) <http://www.urich.edu/~jolt/v1i1/hardy.txt>.

there are other slices missing as well. The second—and perhaps largest—missing slice might be termed the "first to market" slice. It is well understood in many industries that copying is inevitable and uncontrollable. Innovators in such markets therefore structure their business plans so as to make their profits in the period between the time their product reaches the market and the time that their competitors have produced a competing clone product. When free-riding competitors reach market it may become impossible to make a profit on the original product, so the innovators simply plan to move on to the next innovation at that time—and of course plan themselves to copy their competitor's innovations whenever the opportunity arises.

A third missing slice might be termed the "sponsor" slice. Dyson suggested that in some instances artists may find patrons who will sponsor their creative works. This could take a variety of forms; Dyson likely had past systems of aristocratic patronage in mind. However, our current system of support for radio and television programming is of course a "patronage" system of sorts, under which corporate sponsors underwrite the costs of creation in return for advertising space. This is perhaps not an altogether attractive option,[50] but advertising sponsors are already ubiquitous on the Internet.

This list of protection alternatives—which is probably incomplete—suggests that the question of copyright for digital works must be reviewed in a new light. Much of the discussion over the Barlow–Dyson thesis has revolved around the veracity of their claim that copyright is dead or mortally wounded. Little or no consideration has gone to whether their alternatives are desirable alternatives, regardless of the health of the copyright statute. In the online environment described here, such alternatives may serve to lower or circumvent high transaction costs where a strong property right would simply lock buyers and sellers into an impossible situation that leaves no room for bargaining. Barlow and Dyson may then be right in this sense: Copyright will not foster creativity by providing a clear and impenetrable legal bulwark against infringement. Yet it may continue to be a viable tool to foster creativity as a muddy entitlement, by shunting buyers and sellers toward informal bargaining solutions in the majority of cases, toward court in other cases, and by providing a flexible decisional rule in the latter cases.

Cost-Lowering Technology

Two particular types of self-help that may emerge under a muddy entitlement regime on the Internet deserve attention because of their peculiar relationship with muddy cyberspace entitlements. Both may serve to lower or circumvent transaction costs. The first of these approaches uses technological means to facilitate in-

---

50. *See* Margaret Jane Radin, *Property Evolving in Cyberspace*, 15 J.L. & COM. 509, 521–22 (1996) (labeling such a prospect as "dystopic").

tellectual property transactions; the second approach uses organizational structure to accomplish similar purposes.

One result of self-help under muddy rules may be the deployment of copyright management systems as a method to deter unauthorized use of digital works, while facilitating purchase of authorized uses. Such systems might seem to alleviate the transaction costs that favored a muddy entitlement. Yet the availability of such systems does not terminate the need for the muddy rule. Such systems are in essence a method of demarcating and tagging information—a form of technological fencing. As already discussed, such actual fences may or may not prove congruent with the desirable legal demarcation of information—like a painting in a frame, or words between covers, tagged and monitored digital works may contain considerable content that is unoriginal, public domain material, or material subject to fair use. This is one reason current proposals to enforce copyright management schemes are misguided: They would in essence change the technological fence into the legal fence, making the two arbitrarily coterminous.

This suggests that a system of muddy entitlements continues to play a role even if technology solves the search and negotiation cost problem, as one purpose of a muddy legal rule may be to shunt ownership disputes into court, where a third party can consider the beneficial spillover in setting the proper level of access to a work—benefits that would not be internalized or considered by the property holder. Netanel's democratic benefits are among such beneficial externalities. In order for this type of third-party review to take place, the legal standard cannot be mapped onto an arbitrary and absolute technological demarcation; it must instead continue to be calibrated to prompt either acquiesence to access, or litigation over appropriate access.

An additional mechanism for lowering search costs is the collective rights management organization. This approach was recently reviewed by Merges.[51] Merges argued that such voluntary associations constitute a private "contract" into liability rules—that is, that the members of such organizations agree to a set fee for use of their intellectual property. According to Merges, such voluntarily adopted fees, because they are set, monitored, and revised by industry experts, are far more efficient than liability rules enacted by the state.

Merges argued that complete entitlements are necessary to the development of such organizations. However, the collectives have grown up under a wide variety of circumstances. Some, such as ASCAP, have formed under legal regimes where an undivided Calabresi–Melamed property entitlement is the central right; others, such as the Harry Fox Agency, have formed where a liability rule is central; yet others, such as industrial "patent pools" have formed under a regime of divided entitlements. Several of Merges' examples, such as fashion and screenwriting

---

51. *See* Robert P. Merges, *Contracting into Liability Rules: Intellectual Property Rights and Collective Rights Organizations,* 84 CAL. L. REV. 1293 (1996).

guilds, formed without the benefit of any entitlement rule—either property or liability—at all.[52]

If there is a common factor among Merges' examples, it may well be that muddy entitlements, whether of the property or liability persuasion, foster such organizations. Many of Merges' examples, such as the copyright societies, appear to have grown up in the shadow of muddy entitlement standards. Muddy standards are similarly featured in many of the new institutional property examples that Merges draw on. Merges made much of private ordering in Southern California ground water basin management as described by Ostrom.[53] The voluntary ground water allocation organizations coalesced, we are told, as a result of the "vague" entitlements to usage under "hazy" state law principles of capture.[54] In particular, the municipalities sought to avoid repeated litigation to clarify their rights, and so voluntarily formed water basin districts and associations that developed and enforced clear private rules for usage.[55] Thus, it would appear that muddy entitlements may foster the development of organizations that privately clarify the entitlements.

CONCLUSION

The past development of property law, including intellectual property law, shows the variety of legal rules that may be applied in different transactional environments. The discussion here should demonstrate that online transactional environments will be as varied as those found in real space, but will often be attended by high transactional costs. We should therefore be suspicious of arguments that promulgate only one type of allocational rule as desirable in every circumstance. In many situations, clear property entitlements may not be the only rule, or even the optimal rule, for fostering digital works. Just as in real-space transactions, there will be a continuing role for muddy rules in cyberspace.

---

52. *Id.* at 1368–69.

53. *See* ELINOR OSTROM, GOVERNING THE COMMONS: THE EVOLUTION OF INSTITUTIONS FOR COLLECTIVE ACTION (1990).

54. Merges, *supra* note 51 at 1323.

55. *Id.*

12

Designing a Social Protocol:
Lessons Learned from the Platform for Privacy Preferences Project

Lorrie Faith Cranor
AT&T Labs-Research

Joseph Reagle, Jr.
Massachusetts Institute of Technology, World Wide Web Consortium

The Platform for Privacy Preferences Project (P3P), under development by the World Wide Web Consortium (W3C), enables users and Web sites to reach explicit agreements regarding sites' data privacy practices. The options chosen in developing the protocols, grammar, and vocabulary needed for an agreement lead us to a number of generalizations regarding the development of technology designed for "social" purposes.

In this chapter we explain the goals of P3P; discuss the importance of simplicity, layering, and defaults in the development of social protocols; and examine the some-times-difficult relationship between technical and policy decisions in this domain.

INTRODUCTION

The relationship between technical choices and the nontechnical consequences of those choices is inherently difficult. In this chapter we discuss several methods for addressing such "policy" decisions in ways that allow engineers and policymakers to apply the methods of their trade to the questions they are best equipped to solve. Our discussion is motivated by our participation in the P3P, a framework for auto-mated decision making about online privacy; however, it is also relevant to a more general set of problems.

When the relationship between technical and policy choices is ignored, it may lead to unintended and undesirable consequences, or situations in which technologies can be coerced to effect covert policies. We highlight two examples of these situations: a proposal for protecting children from harmful materials online that could have far-reaching unintended consequences for the Internet, and an urban design decision that effected covert policies.

The debate surrounding the Communications Decency Act (CDA) presents several examples of technologies (and proposed technologies) that could lead to unintended consequences. For instance one of the proposals for preventing children from accessing harmful materials online would have required the next version of the Internet protocol to include support for labeling each piece of Internet data with respect to the age of its sender. A note entitled *Enforcing the CDA Improperly May Pervert Internet Architecture* stated that by including such functionality within Internet routers the simplicity, low cost, and radical scalability of the Internet would be jeopardized (Reed, 1996):

> No matter what you believe about the issues raised by the Communications Decency Act, I expect that you will agree that the mechanism to carry out such a discussion or implement a resolution is in the agreements and protocols between end users of the network, not in the groups that design and deploy the internal routers and protocols that they implement. I hope you will join in and make suggestions as to the appropriate process to use to discourage the use of inappropriate architectural changes to the fundamental routing architecture of the net to achieve political policy goals.

An example of a covert mechanism designed to implement social policy is the decision of mid-20th-century New York city planner Robert Moses to design his roads and overpasses so as to exclude the 12-foot-high public transit buses that carried people—often poor or of color—to the parks and beaches he also designed (Winner, 1988). Not only did he fail to separate the mechanism of transportation from social policy, he did so in such a way that his own biases were substituted in place of legitimate policy processes.

Bob Scheifler, a developer of the X Windows system, did recognize the importance of separating mechanism and policy and is often quoted for his useful maxim of "mechanism not policy." The application of this statement was toward the rather technical decision of how applications should avoid setting X resources directly, but instead allow these "policy" decisions to be controlled by the user (Scheifler, 1987). The result was a mechanism that allowed user control over graphical elements and the window system.

In the online realm, protocols have been developed to solve technical problems such as uniquely addressing computers on a network and preventing network bottlenecks. However, new protocols are being developed that are driven by explicit policy requirements. For example, metadata—ways to describe or make statements about other things—and automated negotiation capabilities are being used as the foundation for applications that mimic the social capabilities we have in the real world: capabilities to create rich content, entrust decisions to an agent, make verifiable assertions, create agreements, and develop and manage trust relationships. We characterize this breed of protocols—including P3P—as social protocols (Reagle, 1997). In contrast to technical protocols, which typically serve to facilitate machine-to-machine communications, social protocols often mediate interactions between humans.

In this chapter we discuss some of the issues we confronted during our work as active members of the W3C P3P working groups and the Internet Privacy Working Group (IPWG) vocabulary subcommittee. Many of the lessons learned in the course of P3P can apply to the development of other social protocols such as those designed to facilitate content control, intellectual property rights management, and contract negotiation.

In the following sections we present a brief background of the P3P effort from both policy and technical perspectives. We then examine the issues of simplicity versus sophistication, layering, and defaults to illustrate ways in which nontechnical decisions are implicitly incorporated into or promoted by technology; we also present several options and recommendations for designers to consider when attempting to mitigate contentions between technical and policy concerns.

Before proceeding, we wish to quickly explain what we mean by policy through a simple definition and example. *Mechanism* is how to technically achieve something; *policy* is what one wishes to achieve. For example, P3P is a mechanism for expressing privacy practices; European Union data protection concepts are an example of a policy. In general, the separation of mechanism and policy provides for great flexibility and allows nontechnical decisions to be made by those most qualified to make them.

P3P BACKGROUND

History

As the use of the World Wide Web for electronic commerce has grown, so too have concerns about online privacy. Individuals who send Web sites personal information for online purchases or customization wonder where their information is going and what will be done with it; some even withhold or falsify information when unsure about a site's information practices (Kehoe & Pitkow, 1997). Parents are particularly concerned about Web sites that collect information from their children (Children's Advertising Review Unit, 1997). Although an April 1997 Harris–Westin survey (Louis Harris & Associates & Westin, 1997, p. 3) found that only 5% of Internet users surveyed said they had experienced what they considered to be an invasion of their privacy on the Internet, 53% said they were concerned that "information about which sites they visit will be linked to their email addresses and disclosed to some other person or organization without their knowledge or consent." These concerns, coupled with confusion about the automatic collection of data and its storage on users' hard drives, have prompted legislators and regulators to take a critical look at online privacy issues and motivated companies and organizations to seek technical solutions to online privacy problems.

In 1996, several organizations launched efforts to develop user empowerment approaches to online privacy (Cranor, 1997). The Electronic Frontier Foundation partnered with CommerceNet to create TRUSTe (renamed from eTRUST), a branded system of "trustmarks" designed to inform consumers about the privacy practices of Web sites and to provide assurances that Web sites accurately report these practices (TRUSTe, 1997). At the beginning of 1997 the IPWG was formed. IPWG is an ad-hoc group coordinated by the Center for Democracy and Technology and comprised of a broad cross-section of public interest organizations and private industry engaged in commerce and communication on the Internet. IPWG began developing a framework for policies and technical tools that give users the ability to make choices about the flow of personal information online (Berman & Mulligan, 1997). In May 1997, the W3C launched the P3P to develop recommended specifications for automated privacy discussions (W3C, 1997).

Technical Background

The P3P is intended to allow sites to express their privacy practices and for users to exercise preferences over those practices. If a relationship is developed, subsequent interactions and any resulting data activities are governed by an agreement between the site and the user. After configuring privacy preferences, individuals should be able to seamlessly browse the Internet; their browsing software (user agent) negotiates with Web sites and provides access to sites only when a mutually acceptable agreement can be reached. Any requests from a service to store data on the user's side must comply with any outstanding P3P agreements. P3P efforts focus on how to exchange privacy statements in a flexible and seamless manner. However, the platform may be used in conjunction with other systems, such as TRUSTe, that provide assurances that privacy statements are accurate.

W3C began P3P by designing the overall architecture for the P3P system and a grammar for expressing privacy practices (W3C, 1997). The P3P grammar specifies the types of clauses that comprise P3P statements. A P3P vocabulary is akin to a Platform for Internet Content Selection (PICS) "rating system" (Resnick & Miller, 1996); the vocabulary specifies the specific terms that fit into the P3P grammar. For example, the P3P grammar specifies that P3P statements must include, among other things, clauses describing any data that are to be collected and the practices that apply to those data; a vocabulary includes a list of specific data practices that are valid in a practice clause. Multiple rating systems or vocabularies can be developed and used independently. IPWG is in the process of designing one such vocabulary.

P3P and PICS are both applications of metadata. Whereas PICS provides only for a simple "label" to be used in describing Web content, P3P employs a grammar that allows clauses to be combined to form richer P3P statements.

DESIGN ISSUES FOR SOCIAL PROTOCOLS

Separating technical decisions from policy decisions is laudable, but such separation is not always readily achievable when designing social protocols, as technical and policy decisions often become intertwined. The line between mechanism and policy may be a fuzzy one, and some aspect of the design often falls within the gray area. We explore three themes of social protocol design that are important to P3P: *simplicity versus sophistication, defaults,* and *layers.* We discuss each with respect to separating mechanism from policy, and when such separation is impossible, we offer potential solutions technologists can use to produce good engineering in the face of contentious policy issues.

We address the themes of simplicity, defaults, and layers in separate sections. However, no understanding of one theme can be applied without an understanding of the others. Decisions about how to set defaults and in what layers to address various concerns can impact the overall simplicity of a software tool; indeed layers and defaults can be created specifically to simplify the user experience and to provide sophisticated options for the users who want them.

Simplicity and Sophistication

In early discussions about P3P, its designers considered ideas for elaborate systems that would contain extremely sophisticated and detailed privacy grammars, tools for robust strategic negotiation, automated privacy enforcement (ways to automatically penalize "cheaters"), cryptography, certificate schemes, and more. There is relatively little that cannot be seen within scope on first blush. However, designers must often simplify their elaborate ideas in favor of system designs that can be readily implemented and used.

With P3P—as with other projects—one must strike a balance between sophisticated capabilities and ease of implementation and use. Difficult decisions will always have to be made with respect to defining certain capabilities as out of scope, or unattainable. However, a number of techniques also exist whereby one can enable sophisticated capabilities that are realizable, readily comprehensible, and easy to use. Such techniques include breaking a large system into smaller modules (modularity), designing a system in layers that have varying levels of accessibility to the user (layering), and building a basic system that allows new features or even entire modules or layers to be plugged in later (extensibility). These techniques also prove useful for separating technical and policy decisions.

DESCRIPTIVE VERSUS SUBJECTIVE VOCABULARIES. PICS and P3P have both been designed as basic frameworks to support automated decisionmaking, but neither specifies the details of a decision-making language. PICS allows multiple third parties to provide these details in the form of rating systems; P3P allows these details to be provided in vocabularies. As a result, third parties can design rating systems and vocabularies that are fairly descriptive or very simple. A sophisticated rating system might have 20 variables that users must set with respect to the type of content they wish to see, whereas a simple rating might have a single "thumbs up" variable. Each system has its benefits. The sophisticated rating system provides more information, but requires greater user involvement in its configuration. The simple rating system is quite easy to use, but conveys less information.

When considering P3P, we are presented with a spectrum of options, ranging from fairly descriptive and sophisticated vocabularies over which users must carefully express their preferences, to simple vocabularies with which users defer the expression of their preferences to others. As shown in Figure 12.1, two important factors that may contribute to a vocabulary's complexity are the character of the information—subjective versus descriptive—and the number of variables. The degree to which the variables are interrelated is another important factor.

To draw the line between descriptive and subjective we present the following understanding: Users express preferences over descriptive information in order to reach subjective "opinions" on which their agents act. Rating systems include both descriptive information and subjective opinion about the appropriateness of content. However, subjective systems can be problematic because users may not know if the bias inherent in the system matches their own. (The most common complaint against filtering technologies today is that decisions are opaque; consequently a user may have deferred to biases that would be offensive to the user if known.) Also, from descriptive information one can always derive a new set of "subjective" opinions. If you are told about the content of a site in terms of violence, language, nudity, sex, who paid to produce it, and the intellectual property rights associated with its use, one can make a thumbs-up or thumbs-down decision. Given only someone else's thumbs down, however, one cannot recapture the descriptive information. Once opinions replace descriptions, information is lost.

Of course, designing a purely descriptive system is not an easy task (Martin & Reagle, 1997), and in many cases may not be practical due to the enormous number of factors that an exhaustive description would entail. Indeed, the choice of which categories to include in a system may be in and of itself a subjective decision (Friedman, 1997). Thus we prefer to think of systems as relatively descriptive or subjective, rather than absolutely so. Rating systems designed to describe adult content have been criticized for not being able to distinguish artistic nudity from sexual nudity, a distinction that is inherently subjective. In this case a system that identifies several factors that describe the type of nudity, even if those factors are somewhat subjective, would likely be overall more descriptive—and more transparent—than a system that rates nudity based on some notion of age appropriateness.

Figure 12.1
A Spectrum of Rating Systems and Vocabularies

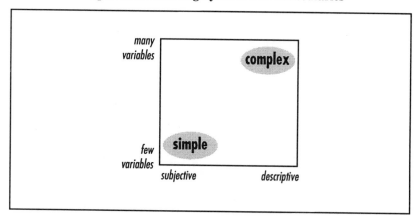

The loss of descriptive information is a significant issue when creating systems for international use where laws and culture vary. Descriptive systems fare best, because a cultural group can always operate on descriptive information, but the biases implicit in a Western "thumbs up" may make such information useless to other cultures. The P3P designers have therefore focused on designing a grammar and vocabulary that allow for the description of how collected information is used rather than subjective statements such as whether information is used in a "responsible" or "appropriate" way.

RECOMMENDED SETTINGS. In the privacy realm both simplicity and sophistication are required. Because people have varying sensitivity toward privacy, we cannot afford to reduce the amount of information expressed to all users to the granularity that is desired by the lowest common denominator—those who want the least information. Fortunately, a complex, descriptive vocabulary can be easily translated into simpler, subjective statements. Rather than manually configuring a user agent using the complex vocabulary, an unsophisticated user can select a trusted source from which to obtain a recommended setting in the form of a "canned" configuration file. These are the settings the user agent will use when browsing the Web on behalf of its user. (A set of recommended settings may be

Figure 12.2

A sample set of recommended settings overlaid on a complex privacy vocabulary. The grids to the right of the settings are shaded to indicate the Practice/Category combinations that are allowed under each setting. Users can select the setting that corresponds most closely to their privacy preferences rather than individually configuring 35 options.

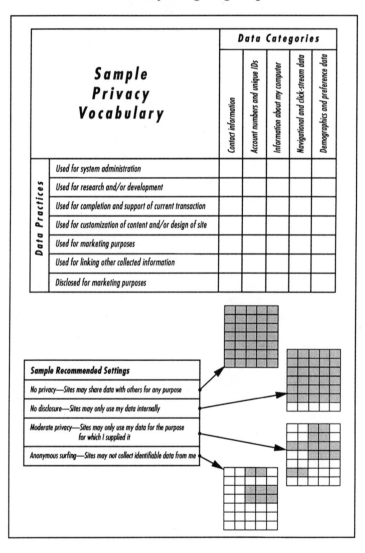

thought of as a subjective vocabulary and simplified grammar, overlaid on top of a descriptive vocabulary; this is an example of layering, a topic we address further in the next section.) Recommended settings capture valuable subjective information that can simplify the user experience while retaining descriptive information in the vocabulary.

Figure 12.2 shows a sample set of recommended settings with a corresponding complex privacy vocabulary. An organization may develop a single recommended setting that reflects its views about privacy, or, as illustrated in the figure, they may develop several settings to provide a simplified menu of options for users. We hope organizations will develop recommended settings that users can download and install in their browsers with the click of a mouse.

This model of referencing others' settings is common in the computer world. For example, it is common for users of the highly customizable Unix operating system to copy the configuration files of more experienced users when first starting out on a new system. In the P3P realm, this model has the added policy benefit of making a distinction between the relatively unbiased provision of descriptive information and those willing to provide recommendations. In such a model, sites would describe their practices in an informative and globally comprehensible vocabulary on which other entities can make recommendations about how users should act. For example, sites may rate with a descriptive vocabulary such as the one illustrated in Figure 12.2, but the user need only choose between a small number of recommended settings. Information is not lost, but the user experience is simplified.

DEFINING A REASONABLE GRAMMAR. We have discussed the use of a descriptive vocabulary over which more subject preferences can be expressed (overlaid). In this section we examine the amount of sophistication appropriate for a grammar and the relationship between the grammar and recommended settings.

At one extreme, we could create a grammar capable of representing any English sentence. However, such a grammar is likely to be difficult to employ in automated discussions, and will often include clauses that prove unnecessary, redundant, or ambiguous in the intended application. It is therefore desirable to restrict the grammar to the number of clauses necessary to make a reasonable set of statements about privacy on the Web. By including extensibility mechanisms, new clauses that are deemed necessary later can be added.

In P3P, the requirements for the grammar are limited to expressing statements about privacy practices. However, determining the universe of statements that might be made about privacy practices is not a simple task. For example, although it seems clear that a privacy grammar must be able to express information about the type of data that might be collected and what the data will be used for, it is unclear whether the grammar must be able to express information about how individuals can update their data or how long the data collector intends to retain the data.

Regardless, a useful check on the adequacy of the P3P grammar is to ask the following two questions. Could a Web site use the grammar (and an appropriate vocabulary) to clearly express that its practices meet the legal requirements for data protection in a given country? Does the grammar provide the ability to express enough information such that a third party (such as a consumer advocate) could issue recommend settings that are meaningful to users?

Layering and User Interface

We already introduced the concept of layering in our discussion of recommended settings, in which we used one layer to simplify or abstract on an underlying layer. This helps us resolve contentions associated with simple versus sophisticated vocabularies. However, all layers are not created equally: Each layer may be owned by a different entity in the technology production chain. Obviously, each owner will be accountable to a different constituency and will have differing policy biases as a result.

The P3P architecture can be modeled as a set of five layers, as shown in Figure 12.3. The P3P layered model serves to both establish a framework for abstraction, and to distinguish the parts of P3P according to the parties responsible for their development. Grammar is the responsibility of the W3C, vocabularies may be defined by IPWG or other organizations, recommended settings may be developed by consumer advocates and other parties trusted by users, and user interfaces will be developed by software implementers. All of these parties contribute toward the final P3P user experience. The entire platform can itself be thought of as a layer (or set of layers) on top of the underlying Internet protocol layers.

This model can help separate policy and technical considerations to some extent, but it is important to remember that decisions made at one layer may have ramifications elsewhere. For example, an overly restrictive P3P grammar may limit the ability of vocabulary designers to create a vocabulary that they consider adequately expressive, or a simplistic user interface may limit the ability of users to express sophisticated preferences that would otherwise be possible with the grammar and vocabulary. Consequently, it is important to determine the most appropriate layer for each decision to be made at and to consider design alternatives that will not overly restrict options at other layers. When considering assigning a policy decision to a given layer, one should consider whether that layer is accessible to the constituency that needs to be represented in making such decisions. The P3P design has been structured so that most of the policy-related decisions can be made in the vocabulary layer—which is being developed by more policy-oriented people than are the other layers. Some of the policy-related decisions will be left to the user interface layer, where it is hoped that developers will either provide the mechanisms necessary to support a broad range of policies or produce a variety of specialized interfaces to address particular consumer needs.

FIGURE 12.3
A five-layer model of the P3P architecture

P3P User Experience	The user experience is the sum of all experiences that the user encouters when using P3P. Such experiences might include the configuration process, and the result of the user's P3P agent exercising the user's P3P preferences. Designers at any level should have in mind the provision of a good user experience.
P3P User Interface	What the user sees on the screen and hears; the user interface can take cues from underlying semantics of the vocabulary but is implementation dependent. It may also provide ways for user to represent themselves as personas.
P3P Recommended Setting	Preferences that are provided by a trusted third party to users who do not wish to configure their own preferences.
P3P Vocabulary	The defined set of words or statements that are allowable in a P3P clause. For example, one vocabulary might define a PRACTICE clause to be either: `{ for system administration}` or `{ for research}` or both.

P3P Grammar	The structure for properly ordering P3P clauses to construct a valid P3P statement. The following example structures clauses (in caps) to make a simple privacy practice statement:
	
	    `For (URI)`
	    `these (PRACTICES) apply for this set of (DATA)`
	
	Note: The above example is much simpler than the actual P3P grammar.

REPRESENTING A CHILD. One issue that comes up in the P3P context is the question of where to represent the concept of a "child." We would like to allow children to browse the Web, preventing them from overriding the preferences set by their parents. We would also like to give parents the option of having their own contact information released rather than their child's information (if information is to be released at all). Need these requirements be internalized in the grammar? If not, should they be referenced in a vocabulary? For example, need the vocabulary be able to reference both the child's and parent's name, or should one create a simple "name" type and associate preferences with it for different personae? (A *persona* is a virtual entity under which one's characteristics and activities can be logically grouped. Some people will have multiple personae; for example, a work persona and a home persona.) For instance, one vocabulary could enable the statement:

```
{we collect your child's name} or
{we collect the parent's name}
```

Another vocabulary could only allow the following:

```
{we collect your name}
```

In the latter case, the distinction between a parent and a child would have to be made outside of the vocabulary—perhaps in the user interface of the browser or operating system.

AUTOMATED DISCUSSIONS ABOUT NEW DATA ELEMENTS. It is important to provide mechanisms that allow users to manage their preferences over data elements, both common elements (such as *name*) and elements encountered occasionaly (such as *nike_shoesize*).

User interfaces may permit users to express separate preferences for each of several categories of data elements. The interface might also allow the user to decide which data elements fall into each category. We can imagine an interface that presents the user with a "bucket" for each category and a set of data elements that can be dragged from bucket to bucket as the user sees fit. The user would then be able to set separate preferences for the elements in each bucket, as shown in Figure 12.4. A user agent with such an interface would be able to engage in automated discussions with Web sites regarding the use of the data elements for which the user has expressed preferences. However, when a site proposes to collect a data element that the user has not expressed a preference about, the user agent will have to prompt the user for input. Although such a system gives users great control, it may cause users to be interrupted with prompts quite frequently.

One of the goals of the project has been, to the extent possible, to maintain a seamless browsing experience on the World Wide Web. If users are interrupted frequently, the experience will not be seamless and it is likely that some users will "swat away" prompts without reading them or grow frustrated and disable the platform completely.

FIGURE 12.4

User agents may allow users to set up buckets for data elements about which they hold similar preferences. The user can drag elements from bucket to bucket and set the preferences on each bucket.

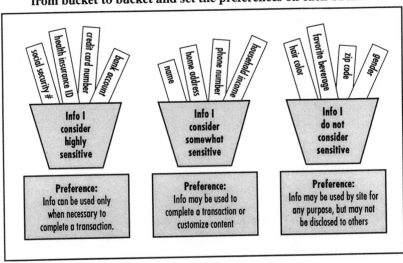

One way to address this problem is to provide elements in a vocabulary that allow Web sites to convey "hints" to the user interface about the nature of each data element. We call these hints data categories. Data categories are characteristics or qualities of data elements. For example, Figure 12.5 shows five data categories: contact information, account numbers and unique IDs, information about my computer, navigational and click-stream data, and demographic and preference data. The data elements "phone number" and "home address" are examples of data that would likely fall into the contact information category. Every time a Web site asks to collect a piece of data it states the category to which the data element belongs. If the user has already expressed a preference about that piece of data or has asked to be prompted about all new data types, the user agent may ignore the data category hint. However, if the user has not expressed a preference about that piece of data, the user agent may use the user's preferences over the data category to proceed in an automated discussion. Consequently, users still retain ultimate control over the data elements and how they are classified, but can also rely on data categories to reduce the number of choices they have to make while browsing.

FIGURE 12.5

User agents may allow users to express preferences about specific data elements as well as general data categories.

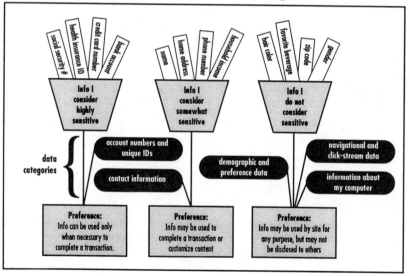

Defaults

The costs associated with configuring software are high in terms of knowledge and time. Consequently, the default settings provided "out of the box" are valuable in that they may determine the online behavior of many users.

Although specifications may avoid defining default values, eventually somebody must make a decision about how to initially set variables, checkboxes, and sliders. We present several approaches that might be taken in establishing defaults. More than one of these approaches may be taken simultaneously for a given piece of software. Each of these options is presented here in contrast to the possibility that implementers could simply create unchangeable settings (implementers decide on some set of default settings and do not allow users to change them). The three options we present are:

1. Leave features that require configuration turned off.
2. Leave features unconfigured, but prompt users to configure them before using the software.
3. Configure features with default settings.

We examine each of these options here and conclude this section with a number of general observations on defaults.

OPTION: LEAVE FEATURES THAT REQUIRE CONFIGURATION TURNED OFF. In this approach, implementers simply deactivate any features that require special configuration. A user who wishes to use these features must explicitly activate and then configure them. This is the approach that was taken by Microsoft for the "Content Advisor" feature in their Internet Explorer 3.0 software. Parents who wished to use this feature to prevent their children from accessing inappropriate content had to explicitly activate it. At that time they could set a password and select a rating system and settings for their child.

Leaving features turned off may be a good alternative when the features are of interest to a small percentage of users or when they may have adverse impacts on users who are unaware of their presence. In the case of P3P, one could argue that setting defaults that would covertly prevent users from accessing sites that do not match the default privacy settings could have adverse impacts on users, and thus P3P should be turned off by default. On the other hand, P3P might be implemented so that its behavior would be obvious to the user, as would be the mechanisms to change the defaults or disable P3P completely.

When features are left turned off, the chances that they will ever be activated are reduced because many users will never take the time to figure out how to activate them (and some will never even realize they exist).

OPTION: LEAVE FEATURES UNCONFIGURED, BUT PROMPT USERS TO CONFIGURE THEM BEFORE USING THE SOFTWARE. In this approach, implementers set up their software so that it is not usable until a user configures it. There are many pieces of software that take this approach for their most important settings, although they generally have defaults for their less important settings that can be reconfigured later.

This alternative has the advantage that users become immediately aware of the existence of features. However, when features take a long time or significant thought to configure, users may grow impatient and give up installing the software or select the simplest configuration options rather than taking the time to select the options they most prefer. This alternative might be useful in P3P if users are initially presented with only a small number of recommended settings.

OPTION: CONFIGURE FEATURES WITH DEFAULT SETTINGS. In this approach implementers decide on some set of default settings for the initial configuration of their software. Users may change these settings later.

This alternative has the advantage that users get the benefits of features without having to take the time to configure them. However, the preconfigured default settings may have a covert impact on the user experience.

In the case of P3P, preconfigured defaults are likely to be controversial, especially in software not advertised as having any particular biases with respect to privacy. A default that provides minimal privacy protections is likely to be criticized by privacy advocates. However, a default that provides strong privacy protections but blocks access to many Web sites is likely to be criticized by people who do not understand why their Web surfing is being interrupted, as well as by the owners of sites that get blocked frequently. On the other hand, there may be a market for products preconfigured to meet a specific need, for example a product preconfigured with strong privacy defaults and explicitly marketed as "privacy friendly."

If implementers select alternatives in which P3P is enabled by default, Web sites that are not P3P compliant may not be viewable without explicit overrides from users. This could be frustrating for users, but might give sites an incentive to adopt P3P technology quickly. (Although to ease the transition, implementers might include tools that would allow users to access sites with no privacy statements if those sites do not actually collect information other than clickstream data.)

When considering the best approach for setting defaults in a particular application, the following ideas should be considered:

- The adoption of a new technology will take time, sometimes a long time. Consequently, choices should not be made that immediately disenfranchise all those who have not adopted the new technology.
- Standards organizations should generally leave default choices to implementers. Implementers are closer to users and the market and will face pressure to configure defaults competitively. For example, privacy-friendly browsers will likely be created if there is consumer demand. (On the other hand, in some domains no effective market exists, and thus the pressure from users will have little weight.)

- It is costly for users to change defaults due to a lack of time or understanding. Efforts should be made at all levels to lower the comprehension and temporal costs associated with configuring technology.
- If preconfigured settings files are available from third parties, users can select from a small menu of options rather than configuring every setting or relying on the manufacturer's preset defaults.

CONCLUSION

Technologists are designing Web applications to address social problems. We characterize such applications as *social protocols* and believe that the good engineering principles of modularity, extensibility, and the development of mechanism—rather than policy—continue to be important in designing these protocols. A unique characteristic of these protocols is that social concerns may explicitly contribute to the requirements of the design. However if collecting and comprehending requirements from a client in the commercial or technical world is difficult, it is much harder to understand the requirements of a society, or multiple societies! Consequently, engineers must be doubly vigilant in the pursuit of mechanism and in promotion of technologies that support multiple policies. For example, a privacy protocol that hard-wired the biases of a few American engineers would be of limited utility in a global Web space.

However, as hard as one may try to draw a line between mechanism and policy, that line is a fuzzy one and some aspect of the design often falls within the gray area. In this chapter we have presented three themes (simplicity vs. sophistication, layering, and defaults) that are useful for navigating through the gray area between mechanism and policy. We have presented our current understanding of these themes as applied to the development of P3P and we have presented recommendations for applying relevant design principles in an explicit and reasoned light. Our purpose has been to convey our own experiences, generalize them so that they may be useful in other applications, and to hopefully inform policymakers about technology development relevant to their own activities.

ACKNOWLEDGMENTS

We thank the P3P working group participants, the members of the IPWG vocabulary subcommittee, and our colleagues at AT&T Labs-Research and the W3C. Many of the concepts discussed herein were not the result of any single individual's effort, but the combined efforts of a number of people. These ideas were developed through many conference calls, e-mail discussions, and face-to-face meetings.

This chapter references work in progress by several organizations. Nothing in this chapter should be construed as representing the final status of any effort.

REFERENCES

Berman, J., & Mulligan, D. (1997, April 15). *Comments of Internet Privacy Working Group concerning consumer on-line privacy* (Report No. P954807, submitted to the Federal Trade Commission, Workshop on Consumer Privacy). Available: http://www.ftc.gov/bcp/privacy/wkshp97/comments2/ipwg049.htm.

Children's Advertising Review Unit of Better Business Bureaus, Inc. (1997, July 14). FTC: *Consumer privacy comments concerning the Childrens Advertising Review Unit* (Report No. P954807, supplement to comment #008, submitted to the Federal Trade Commission, Workshop on Consumer Privacy). Available: http://www.ftc.gov/bcp/privacy/wkshp97/comments3/CARU.1.htin.

Cranor, L. (1997). The role of technology in self-regulatory privacy regimes. In *Privacy and self regulation in the information age*. Washington, DC: U.S. Department of Commerce, National Telecommunications and Infrastructure Administration. Available: http://www.ntia.doc.gov/reports/privacy/selfreg5.htm#5B.

Friedman, B. (1997). *Human values and the design of computer technology*. Cambridge, UK: Cambridge University Press.

Kehoe, C., & Pitkow, J. (1997, June). GVU's seventh WWW user survey. Available: http://www.gvu.gatech.edu/user_surveys/survey-1997-04/.

Louis Harris & Associates, & Westin, A. F. (1997). Commerce, communication and privacy online: A national survey of computer users (Conducted for Privacy & American Business). Hackensack, NJ: Privacy & American Business.

Martin, C. D., & Reagle, J. M. (1997). A technical alternative to government regulation and censorship: Content advisory systems for the Internet. *Cardozo Arts & Entertainment Law Journal, 15*(2), 409–427.

Reagle, J. (1997, January). *Social protocols: Enabling sophisticated commerce on the Web*. Presentation at the Interactive Multimedia Association, Electronic Commerce For Content II. Available: http://www.w3.org/Talks/9704-IMA/.

Reed, D. (1996, April 28) In Cipher Electronic Issue #14. Available: http://www.itd.nrl.navy.mil/ITD/5540/ieee/cipher/old-conf-rep/comment-reed-960331.html.

Resnick, P., & Miller, J. (1996, October). PICS: Internet access controls without censorship. Communications of the ACM, *39*(10), 87–93.

Scheifler, R. (1987, June). RFC 1013 X WINDOW SYSTEM PROTOCOL, VERSION 11. Available: ftp://ds.internic.net/rfc/rfc1013.txt.

TRUSTe. (1997, June.) The TRUSTe story. Available: http://www.truste.org/program/story.html.

Winner, L. (1988, January). The whale and the reactor: A search for limits in an age of high technology. Chicago: University of Chicago Press.

World Wide Web Consortium. (1997, December). *Platform for Privacy Preferences Project* [Web site containing overview and status of P3P. W3C recommendations with regards to P3P will be posted here when they become available publicly]. Available: http://www.w3.org/P3P/.

V

COMPARATIVE STUDIES IN TELEPHONY AND SATELLITE POLICY

The Paradox of Ubiquity:
Communication Satellite Policies in Asia

Heather E. Hudson
University of San Francisco

INTRODUCTION

Asia's vast land area, its population of more than 3 billion, a teledensity of fewer than 3 lines per 100 inhabitants in the lower income countries, and its unserved rural areas and isolated islands of Southeast Asia and the Pacific all make the Asia-Pacific region attractive for satellite communications. In fact, the Asia-Pacific region now lags behind only North America in number of commercial satellites in orbit, and leads the world in the number of satellites on order (Boeke & Fernandez, 1997).

The race to add capacity involves international systems such as Intelsat and PanAmSat, regional systems, and domestic satellites, many of which are designed with regional coverage. In a few years, the Asia-Pacific region has moved from a shortage of satellite capacity to a potential excess: "In fact, so many regional birds are being launched that a glut of capacity exists, driving down prices for end users. This scenario could lead to consolidation or business losses for operators in the region" (Boeke & Fernandez, 1997, p. 38). Yet not all of the potential customers in the region may benefit from the burgeoning number of satellites, as restrictive policies and lack of terrestrial infrastructure may preclude access to new satellite services.

This chapter examines the issues surrounding the increase in satellite capacity serving the developing countries and emerging economies of the Asia-Pacific region, and derives lessons for satellite policymaking in other regions. Issues addressed include:

- Competition vs. monopoly: Regional and international systems such as Asiasat, Apstar, and PanAmSat challenge not only Intelsat, but national broadcasters who are realizing that they must change their format and content in the face of the invasion of foreign programming.

- Access vs. control: High-powered satellites ("hot birds") that can transmit to wok-sized antennas make efforts to control access to foreign content largely futile. Also, satellite broadcasters are using digital technology to precensor programs, or to tailor them for different markets.
- Nationalism vs. regionalism: Many Southeast Asian nations have invested in satellites ostensibly to reach their rural and isolated populations. Although satellites can be very cost effective to reach such unserved areas, most countries need at most several transponders rather than a dedicated domestic satellite. To generate more revenue, the footprints of their satellites are designed to cover neighboring countries, so that many countries are covered by multiple competing "national" satellite systems.
- Flags of convenience: Countries are trading orbital slots to investors in return for concessions that may not be sufficient compensation to provide communications to their own underserved populations.
- The last mile problem: Poorer countries still lack the basic infrastructure that will be needed to extend satellite services in rural areas.

THE PROLIFERATION OF INTERNATIONAL AND REGIONAL SATELLITES

Intelsat continues to provide global and regional connectivity, but has also developed new strategies to serve Asian nations. In addition to providing more capacity, higher power, and greater connectivity on its Intelsat VIII series, Intelsat has negotiated with China's ChinaSat for the use of half of the transponders on ChinaSat 5 to give additional C-band coverage of the region and has leased 11 C-band transponders on India's 2-E for additional capacity over the Indian subcontinent. Intelsat has also created an additional Asia-Pacific customer service area with an initial satellite located at 91.5 degrees East (*APT Yearbook*, 1996). PanAmSat achieved global connectivity including coverage of Asia and Africa in 1995. Its Pacific and Indian Ocean satellites cover the region, and provide links to Europe and North America. PAS-4, which covers South Asia and the Indian subcontinent, now carries numerous international television signals on its South Asia/Middle East beam (*APT Yearbook*, 1996).

Asiasat 1, launched in 1990, opened a new era in Asian telecommunications, offering high-powered capacity for television distribution, data networking, and domestic telephony. Based in Hong Kong, Asiasat is owned by Cable and Wireless, China International Trust and Investment Corporation (CITIC), and Hutchison Whampoa. Television broadcasters from China, Mongolia, Pakistan, and Myanmar (Burma) use Asiasat for domestic distribution. However, Asiasat's most significant impact has been through its distribution of News Corporation's Star TV, a group of commercial channels designed for pan-Asian audiences. Asiasat 2, launched in 1995, covers 53 countries including Asia, the Middle East, Australia, and Eastern Europe.

APT's Apstar, also based in Hong Kong, became Asiasat's first major competitor with the launch of Apstar-1 in 1994. Apstar-1's footprint reaches from Korea and Japan to Indonesia and west to eastern Russia. Apstar 1A includes coverage of India. Apstar 2 was lost in a Chinese launch failure in 1995, but is being replaced by Apstar 2R, with a total of 44 transponders and coverage of Asia and Australia and parts of Europe and Africa. Orion also plans to provide Asia-Pacific coverage with a hybrid satellite equipped with 10 C-band and 33 Ku-band transponders located at 139 degrees East longitude.

NATIONAL SATELLITE SYSTEMS

Satellites have many advantages for countries such as Indonesia and the Philippines, with archipelagoes consisting of thousands of islands, and for countries such as Vietnam and Laos with very limited infrastructure and unserved rural and remote communities. However, there is not necessarily enough demand in each developing country to justify its own domestic satellite system.

Nevertheless, Indonesia, India, and China are being joined by new members of the national satellite club including Thailand, Malaysia, South Korea, the Philippines, and Laos (see Table 13.1). These national systems are being designed with regional beams so that they can provide services to their neighbors, but the neighbors seem intent on procuring their own satellites. As noted earlier, capacity is also available from regional and international systems including Asiasat, Apstar, and PanAmSat, as well as from Intelsat.

TABLE 13.1

Developing Countries' National Satellite Systems in the Asia-Pacific Region

Country	System	Date of First Launch
Indonesia	Palapa	1976
India	Insat	1982
China	Chinasat	1988
Thailand	Thaicom	1993
South Korea	Koreasat	1993
Malaysia	Measat	1996
Philippines	Mabuhay*	1997
Laos	Lstar	1998 (planned)

* Agila I was purchased from Indonesia and moved to cover the Philippines in 1996.

Indonesia

In 1976, Indonesia became the first developing country with its own domestic satellite system, with the launching of Palapa 1, a Hughes-built satellite similar in design to Canada's Anik and the U.S. Westar satellites, also built by Hughes. Indonesia's second-generation satellites cover both its own archipelago and the neighboring ASEAN[1] nations. Its third-generation Palapa C series provide coverage from Iran to Vladivostok and south to Australia and New Zealand.

In 1993, state-owned PT Telekom and PT Indosat established a joint venture called PT Satelit Palapa Indonesia (PT Satelindo) with privately owned PT Bmiagraha Telekomindo to operate the Palapa system. In 1995, De Te Mobil, a subsidiary of Deutsche Telekom, bought a 25% stake in Satelindo. Another consortium called Indostar comprising state-owned and private companies is investing in four small satellites designed for domestic direct-to-home digital TV and audio services (Hudson, 1997).

India

India began experimenting with satellite communications for rural development in the 1970s, when it borrowed NASA's ATS-6 satellite to transmit educational television programs to community receivers in rural villages (ATS-6 was moved to cover India; it was later repositioned over the United States). India's first domestic satellite was launched in 1982. The Insat system is now in its second generation, providing television distribution, long distance telecommunications, data relay, and meteorological earth observation. To date, all transponders are provided without charge to state agencies; the Department of Space is considering leasing transponders to private users (*APT Yearbook*, 1996).

China

China began using satellites for domestic communications in 1985 when it leased Intelsat capacity to link the remote centers of Urumqi, Llasa, and Hohhot with Beijing and Guanzhou. Intelsat is now used to link other remote locations and to transmit China Central Television (CCTV) as well as for private VSAT (very small aperture terminal) networks. In 1988, China launched two domestic satellites that have been used to transmit television and FM radio programs as well as voice and data. A VSAT network provides voice communications for Tibet and for remote provinces. The Tibetan network first used Asiasat-1, but has been transferred to Apstar-1 (*APT Yearbook*, 1996).

---

1. Association of South East Asian Nations; members currently include Brunei, Indonesia, Laos, Malaysia, Myanmar, Singapore, Philippines, Thailand, and Vietnam.

Thailand

Thailand's first domestic satellites, Thaicom 1 and 2, were launched in 1993 and 1994, and are owned and operated by Shinawatra Public Satellite Company, which has a monopoly on satellite operations in Thailand until 1999. Thaicom 1 and 2 are used for national and cable television distribution, educational television, VSAT data networks, and rural telephony. The next generation, beginning with Thaicom 3, launched in April 1997, has C-band footprints that cover Asia and parts of Australia, Europe, and Africa, and a regional footprint from India to Indochina. Its high-powered Ku-band transponders are designed for direct-to-home television distribution for Thailand and surrounding regions with beams covering Thailand and India, as well as switchable beams.

South Korea

Koreasat-1 was launched in 1995, but did not reach geostationary orbit during the launch. Onboard thrusters were used to boost the satellite into orbit, but at an expense of 4.5 years of its projected 10-year life. A second satellite has been colocated at 116 degrees E to provide additional capacity. The Koreasat system has high-powered Ku-band capacity for direct-to-home services and high-definition television as well as private VSAT data networks and public voice and data services.

Malaysia

Malaysia's satellite system, Measat, began operation in 1996. Measat is a privately owned venture of Binariang, which also operates domestic fixed and cellular networks. Beams cover the ASEAN nations as well as southern India, and as far as northern Australia and southern China. The system is used primarily for direct-to-home pay television, as well as for distribution of Radio-TV Malaysia and for VSAT networks.

The Philippines

The Philippines has been using satellite technology for domestic services by leasing capacity on Indonesia's Palapa system. In 1995, a consortium of 17 companies called Philippine Agila Satellite, Inc. (PASI) announced plans for a national system. However, the largest member, Philippine Long Distance Company (PLDT), left the group and founded another consortium called Mabuhay Philippine Satellite Corporation (MPSC), whose members include several domestic telecommunications and broadcasting companies as well as Indonesia's PT Pacifik Satelit Nusantara and China's Everbright Group (Dance, 1997).

President Ramos stated that he wanted a Filipino satellite in orbit in time for the Asia-Pacific Economic Council (APEC) meeting held in the Philippines in November 1996. MPSC managed to comply by obtaining Indonesia's B2P satellite (now called Agila I) and moving it into a slot at 144 degrees East obtained from the operators of Japan's Superbird. Its own Mabuhay satellite was launched in 1997.

THE NATIONAL FLAG CARRIER SYNDROME

The national flag carrier syndrome that seems to require every country to have its own airline appears to have mutated to satellite systems in the Asia-Pacific region. Despite the introduction of competition in international and regional systems with beams shaped for domestic and subregional coverage, Asia's developing countries appear determined to have their own satellite systems. The symbolic value of a national satellite system is reflected in a trade press article on Filipino satellite initiatives entitled "The Philippines Comes of Age" (Dance, 1997).

This sentiment can be traced back more than two decades to the origins of the Palapa system. President Suharto wanted Palapa to be launched before the 1976 elections; design and construction were constrained to meet this deadline (Hudson, 1990). The satellite policy environment in the Philippines in 1996 was remarkably similar to the situation in Indonesia in 1976. President Ramos stated that he wanted a Filipino satellite in orbit in time for the APEC[2] meeting held at Subic Bay (the former U.S. Navy base) in the Philippines in November 1996. MPSC, the PLDT-led consortium, was able to comply, ironically through its Indonesian member, who was able to arrange acquisition of both an Indonesian satellite and an orbital location.

From Domsat to National Flag Carrier

In order to justify investment in a satellite system, countries are shifting their focus from strictly domestic coverage to regional and even intercontinental markets. Again, Indonesia was the pioneer. As noted earlier, although the first Palapa was designed specifically for Indonesia, later satellites included coverage of the ASEAN region; Malaysia and the Philippines were major customers. It may have been Indonesia's success in attracting foreign customers that encouraged other ASEAN countries to invest in their own satellites.

An argument given for investing in a Filipino satellite was that it would help reduce the national balance of payments deficit at a faster pace because scarce foreign exchange would no longer be needed for foreign transponder rental (*APT*

---

2. Asia-Pacific Economic Community, comprised of economies (a designation chosen to allow inclusion of Hong Kong and Taiwan) in Asia, the Pacific Basin, and the Pacific Rim.

Yearbook, 1996). Yet these systems themselves represent an enormous investment of hard currency, as domestic satellite systems (including the satellite, launch, insurance, and master control stations) typically cost at least $200 million. During the APEC summit, the chairman of the U.S. Export–Import Bank signed a letter of intent with MPSC for up to $225 million in direct financing for Agila II, later renamed Mabuhay (Dance, 1997). Revenue for transponder leases from foreign customers could, however, provide foreign currency to repay loans for building and launching the satellites.

With coverage on Thaicom 3 of the Middle East and parts of Africa, Thailand's Shinawatra will now compete with global operators Intelsat and PanAmSat, as well as potentially Arabsat, in addition to regional systems such as Asiasat and Apstar. New generations of Palapa reach from Iran to Australia and New Zealand, and Measat has a similar reach and offers spot beams over India and the Philippines.

Flags of Convenience

The International Telecommunication Union (ITU) allocates satellite orbital locations to nations, not to operators. Thus, in order to obtain slots, operators must turn to national governments to act on their behalf. Governments, in turn, may seek to obtain benefits in return for use of their slots. The Asia-Pacific region has pioneered an approach that could be called "flying flags of convenience," similar to the practice of shipping companies that register their fleet in countries with favorable regulations and tax policies. Countries too small or poor to have their own satellites can file with the ITU for slots that they can then turn over to satellite investors in return for a negotiated compensation.

Foreign investors who anticipated that trade in orbital slots would be a good business persuaded Tongan officials to sanction Tongasat, an entity that filed for slots on behalf of Tonga and then leased them to satellite operators. Although the principals in Tongasat may have benefited financially, it appears that the people of Tonga have not benefited from improved telecommunications, despite their rural teledensity of only about 5 lines per 100 inhabitants, and the isolation of their northern islands, which could be ideally served by satellite.

The apparent lack of policy linking the filing for orbital slots with the utilization of satellites to improve Tongan telecommunications can be seen in entries on Tonga juxtaposed in the *1996 APT Yearbook*. Under "Government Policy," the *Yearbook* states:

An important condition to fulfil the regional development objects of the Tonga government is to ensure the provision of appropriate telecommunications services to the isolated areas...Because of the long distances involved and the low population density, particularly in the Ha'apai and Vava'u Groups, the networks on these isolated scattered islands are invariably much more expensive to install than those in Nuku'alofa, the capital. (*APT Yearbook*, 1996, p. 556)

Under "Satellite Systems," the *Yearbook* adds:

To date, two companies have contracted with Tongasat to place satellites into Tonga's geostationary slots. These satellites will cover all of the Asia-Pacific region and will have footprints that can extend to Europe and the Middle East and the East Coast of the USA. (*APT Yearbook*, 1996, p. 557)

But apparently these satellites will not to link Ha'apai and Vava'u to the rest of the country, nor Tonga to other countries in the region.

Laos has allowed a foreign operator to use its slot in return for some compensation. The Lstar satellite, built by Space Systems/Loral, is ostensibly a Laotian satellite, but is actually a venture of the Thai Asian Broadcasting and Communications Network Company (ABCN), operating under a 30-year concession from the Lao PDR government (*APT Yearbook*, 1996). With a national teledensity of only .38, rural teledensity of .18, and a highly rural population, Laos would appear an ideal candidate to benefit from use of a satellite to close its communications gaps. However, in return for a Laotian orbital slot, the Thai investors in the system, who plan to use it primarily for television distribution, have allocated only one transponder for Laotian telecommunications. Although digital compression can expand the number of voice and data circuits that can be squeezed onto one transponder, capacity will be very limited, especially if the Laotians plan for television distribution as well as telephony. Also, the very limited existing Laotian infrastructure will require additional investment, such as wireless links to neighboring villages, to extend service from each rural satellite terminal.

The Philippine government also hopes to benefit from concessions in return for orbital slots obtained for private operators. When the initial PASI consortium was established, the government was to secure four orbital slots and associated frequencies from the ITU. In exchange for the slots the consortium was to allot the government one free transponder for every 12 available transponders on the national satellite. As noted earlier, PLDT left PASI, and its new group, MPSC, moved Palapa B2P to 144 degrees East, and contracted for construction and launch of a second satellite. In the meantime, the government acquired 153 and 161 degrees East; its policy in mid-1997 was to let MPSC use 153 degrees East in return for the original guarantee of free transponders; other interested parties including PASI would have to bid on the slots (Dance, 1997). Under this policy, two transponders on Mabuhay 1 are available for government and public service use.

THE LAST MILE PROBLEM

Satellites are generally an ideal means of serving isolated populations, but to extend service beyond a public telephone colocated with the satellite terminal requires installation of terrestrial infrastructure, for example, to connect other customers in the community or to link the station with surrounding villages. At present, very little rural infrastructure exists in many developing countries of the Asia-Pacific region. (see Table 13.2).

TABLE 13.2
Rural Teledensity in Selected Developing Asian Nations

Country	Teledensity 1995	Teledensity Outside largest city: 1995
Cambodia	1.01	0.05
China	3.35	3.23
India	1.07	0.95
Indonesia	2.10	0.20
Laos	0.56	0.07
Malaysia	18.32	14.73
Myanmar (Burma)	0.38	0.18
Philippines	2.58	0.95
Thailand	7.66	2.94
Vietnam	1.58	0.71

Derived from ITU, *World Telecommunication Development Report* 1996/97, Geneva: International Telecommunication Union, 1997; and ITU, Asia-Pacific *Telecommunication Indicators 1997*, Geneva: International Telecommunication Union, 1997.

Of course, lack of infrastructure is a problem that satellite systems are designed to solve, but additional investment is required to build out the network to connect other customers in a village or to link neighboring villages. The existence of rural satellite facilities can provide the incentive to invest in local facilities because of the revenue generated by using the satellite network. For example, in Alaska, several rural telephone cooperatives were established to provide local telephone service after satellite earth stations were installed in the 1970s. However, two factors were important stimulants for this investment: First, the settlement agreements between the "high-cost" Alaskan rural phone companies and the long distance carriers enabled the rural cooperatives to keep most of the revenue generated by calls originating and terminating in the villages. Second, rural cooperatives could apply for federal Rural Electrification Administration (REA)[3] loans to build or up-

---

3. Now called the Rural Utilities Service (RUS).

grade their local facilities. The REA program provided loans at below market interest rates and access to technical assistance on design, installation, operation, and maintenance of rural telephone systems.

Yet the existence of a domestic satellite system does not seem to have stimulated investment in rural terrestrial "last mile" networks in Indonesia or India, which have had domestic satellite systems since 1976 and 1982, respectively. Palapa was named for the Palapa oath, made by Gjah Mada of the ancient Majapahit Kingdom, who vowed that he would to continue to work until all the islands of the archipelago were united (*APT Yearbook*, 1996). The government has set an ambitious target of adding 5 million lines between 1994 and 1999, which would raise teledensity to about 2.8. Rural teledensity in 1995 was only 0.2 lines per 100, or 2 lines per 1,000 rural Indonesians. Thus, although the Palapa system has been in existence for more than 20 years, it has not linked most Indonesians in the archipelago. Similarly, India, whose Insat system is almost 15 years old, has a rural teledensity of less than 1 telephone per 100 inhabitants. Insat alone has not been able to bridge the rural communications gap.

India lags behind the average of all low-income countries in both teledensity and compound annual growth rate (CAGR) of telephone lines (see Table 13.3). Indonesia still lags far behind the average of all lower middle income countries, but its CAGR is an impressive 14.8%. The critical factor appears to be the restructuring of the telecommunications sector in Indonesia, which has created new opportunities and incentives to extend services throughout the country. Indonesia has begun to open basic telecommunications to private investment through incentive plans such as Build Operate Transfer (BOT) and Build Operate Own (BOO). India's restructuring, involving licensing of second local operators, has become bogged down in political upheavals and allegations of corruption. However, investment incentives in Indonesia, Thailand, and the Philippines, which requires installation of rural lines in return for gateway and urban cellular licenses, may spur the rural investment needed to take advantage of the satellite links.

TABLE 13.3
India and Indonesia: Change in Teledensity, 1984–1994

	Teledensity		
	1984	*1994*	*CAGR**
India	0.39	1.07	10.7%
Average of all lower-income countries	0.36	1.48	15.1%
Indonesia	0.33	1.33	14.8%
Average of all lower middle income countries	4.51	8.40	6.6%

* Compound Annual Growth Rate. Derived from: ITU, *World Telecommunication Development Report 1996/97*, Geneva: International Telecommunication Union, 1997.

Yet even this restructuring may not result in utilization of available satellite capacity. For example, many countries have adopted the "carrier of last resort" model, designating one carrier (typically the former post, telephone and telegraph (PTT) or national monopoly carrier) with the responsibility of serving rural areas if no other provider offers service. If the carrier of last resort owns or has a major investment in the domestic satellite system, it would have an incentive to use the satellite. If not, it may choose instead to install its own long distance network, even if this choice requires a much larger capital outlay. By installing its own network, it can depreciate its own equipment and retain more of the revenues generated. Thus, for example, in Malaysia, Telekom Malaysia is planning a microwave network for remote areas of Sarawak and Sabah, instead of using capacity on the privately owned Measat domestic satellite.[4]

SATELLITES AS TROJAN HORSES

Asian satellites have served as "electronic Trojan horses," introducing both new services and new operators that governments were reluctant to condone. The most visible example is foreign television content, first introduced by Star TV, News Corporation's package of commercial programming carried on Asiasat. Star now estimates that it reaches 53 million viewers from Israel to Taiwan (Hudson, 1997). Although India's Insat had an eight-year head start on Asiasat, its national TV network, Doordarshan, lost viewers to Star TV. There are now an estimated 900,000 TVRO (TV receive only) antennas in India, but most of them are used to receive programming from Asiasat rather than Insat. India also has an estimated 40,000 cable TV operators (Waller, 1997), an industry that was born as a means to redistribute StarTV, and now offers more than 45 satellite channels. In response to Star TV, Doordarshan has revamped its programming, adding more films and soap operas, and a channel aimed at wooing the elite back from foreign channels. Meanwhile, Measat, Thaicom, and other satellite operators have aimed beams at India in anticipation of distributing both foreign and domestic commercial television to Indian viewers. India's Modi Entertainment Network will offer a commercial direct-to-home package on Thaicom's India beam (Waller, 1997).

National broadcasters in other Asian countries are also revamping their program schedules to lure viewers back from StarTV and other imported satellite channels. Some countries have gone further and attempted to prevent reception of foreign

---

4. Both Canada and Australia have invested in domestic satellite systems, but still rely on terrestrial networks for many rural and remote areas. Bell Canada had no incentive to use Telesat Canada's Anik system in the 1970s, and chose to build microwave networks in northern Ontario, for example, using Anik satellites only to reach the mid- and high Arctic. Telecom Australia (now Telstra) chose to build a microwave network to link isolated outback communities in the 1980s rather than using the domestic Aussat system.

channels by banning TVROs. Singapore and Iran are among the nations that ban TVROs. Other countries such as China, Malaysia, and Indonesia officially allow reception of only domestic programming. To satisfy demand for more television channels, Singapore and China are encouraging construction of cable television systems. The advantage of cable, in addition to controlling access to generate revenue, is that content can be monitored and controlled. For example, Chinese cable viewers were unable to view CNN coverage of the June 4, 1997, rallies in Hong Kong in remembrance of Tiananmen Square.

Satellites also provide a means to bypass the public telecommunications network, regardless of whether bypass has been officially sanctioned by national regulators. Although little publicized, these VSAT networks linking businesses and government agencies may generate the greatest economic benefits of all the satellite-based services. In 1993, China's Ministry of Posts and Telecommunications (MPT) deregulated VSAT data networks, spurring implementation of private VSAT networks and competition in VSAT services. Apart from the MPT, several other state agencies have their own VSAT networks, including the Bank of China, stock exchanges, Xinhua News Agency, People's Daily, Customs, Civil Aviation, Ministry of Communications, Ministry of Energy, and the People's Liberation Army (*APT Yearbook*, 1996).

India's first packet-switched satellite-based data network, known as Nicnet, was installed by the National Informatics Centre (NIC), a government agency that was able to make the case that it needed its own satellite network to transfer data because the public network was so limited and unreliable. The NIC now has 700 VSATs linking 500 centers; in 1994, it added high-speed earth stations linking 14 cities. Its latest project offers satellite-based Internet access using Hughes DirecPC terminals for government agencies, government-sponsored research and education units, and other government-funded bodies ("Internet Boost for NICSI," 1997). Another innovative user of Indian VSAT networks is the National Stock Exchange (NSE), which has established a satellite-based trading system. The newest of India's 23 stock exchanges, the NSE began trading in 1994. By mid-1995, more than 225 members had been linked via satellite to the NSE's mainframe computer, enabling them to view online market information, place orders, and execute trades directly from their offices (Narain, 1996). Yet India's public telecommunications system remains mired in inefficiency, with plans to allow competing operators not yet implemented.

SATELLITES: THE INVISIBLE INFRASTRUCTURE?

As noted earlier, many developing countries in the Asia-Pacific region have invested in satellites to deliver television and to provide voice and data services. Others use Intelsat, PanAmSat, and regional satellites for domestic communications. Yet satellites do not seem to be considered part of the information infrastructure being planned under Asian or Asia/Pacific Information Infrastructure (AII or APII) initiatives. Most discussion of AII in the region focuses on extension of optical fiber networks for Internet access and broadband services such as asynchronous transfer mode (ATM). Malaysia promotes its Multimedia Supercorridor near Kuala Lumpur as the showpiece of its Vision 2020 strategy; Indonesia cites its microwave and fiber optic backbone linking major cities. At a meeting of APT representatives in Bangkok on information infrastructure in June 1997, national and regional satellites were almost completely ignored in the review of plans for AII. Yet satellites have enormous potential to provide not only basic telephony but also capacity for Internet access, distance education, telemedicine, and teleworking in rural and remote areas of the Asia-Pacific region.

LESSONS FROM THE ASIA-PACIFIC EXPERIENCE

Of course, the experience in the Asia-Pacific region is not unique; other developing countries have faced similar obstacles to maximizing benefits from their investment in satellite systems. However, the Asia-Pacific region has the longest experience with national systems, and the greatest number of national and regional satellites outside North America. Lessons from its two decades of experience that may be valuable for future satellite endeavors in the Asia-Pacific region and in other developing regions include:

- **Incentives for investment:** Perhaps most important, investment in satellites alone will not overcome communications gaps. There must be investment in terrestrial infrastructure, and incentives to extend this infrastructure in rural or impoverished areas that are not highly attractive to investors.
- **Incentives for usage**: In addition, there must be incentives to use the satellite system. Even where there is demonstrable unmet demand, a satellite can soon become a white elephant if operators have no incentive to use it, but may choose instead to build their own terrestrial networks, even if these are more costly.
- **Regulatory reform:** Third, the regulatory environment must facilitate satellite operations, for example through legalization of bypass and resale of satellite capacity, and requirements for interconnection agreements between satellite and terrestrial carriers.
- **Starting with users**: Finally, as with all telecommunications planning, it is important to understand the customer. Satellite operators that have recognized unmet demand for video and data services have been successful through responding to user needs Among the factors to consider are service requirements, pricing, and communities of interest (i.e., where users need to communicate). These factors may be particularly relevant for the new Global Mobile Personal Communication Systems (GMPCS), includ-

ing both low-earth-orbiting (LEO) and geostationary satellite systems, that plan coverage of the Asia-Pacific region. Most of these systems plan to offer maximum bandwidth of 9600 bps. Given the growth of the Internet in Asia, there is likely to be significant demand for higher capacity to surf the Internet. Also, a satellite system that links a rural field worker to a national head office only by transmitting through the terrestrial networks of another country (such as ICO Global Communications) is not likely to be able to offer affordable or even perhaps reliable service. And a system that cannot withstand heavy rain, humidity, or dust (such as high-powered Ku-band and Ka-band geostationary systems) will not deliver the reliable services that users will come to expect.

This analysis demonstrates that the extension of reliable telecommunications to developing regions of Asia and the Pacific requires more than investment in satellite technology. As Arthur C. Clarke, the first to propose geostationary satellites, observed in 1983: "We have now reached the stage when virtually anything we want to do in the field of communications is possible. The constraints are no longer technical, but economic, legal, or political" (cited in Hudson, 1990).

ACKNOWLEDGEMENTS

The author conducted research for this chapter with the assistance of the Fulbright Foundation, which awarded her an Asia-Pacific Distinguished Lectureship to visit Singapore, Malaysia, Hong Kong, and Japan.

REFERENCES

APT Yearbook 1996. (1996). Bangkok, Thailand: Asia-Pacific Telecommunity.

Boeke, C., & Fernandez, R., (1997, July). Via Satellite's global satellite survey. *Via Satellite*, pp. 32–38.

Dance, T. (1997, February). The Philippines comes of age. *Asia-Pacific Satellite*, pp. 23–25.

Hudson, H. E. (1990). *Communication satellites: Their development and impact.* New York: The Free Press.

Hudson, H. E. (1997). *Global connections: International telecommunications infrastructure and policy.* New York: Van Nostrand Reinhold.

Internet boost for NICSI. (1997, June). *Asia-Pacific Satellite*, p. 4.

International Telecommunication Union. (1997a). *Asia-Pacific telecommunication indicators 1997.* Geneva: Author.

International Telecommunication Union. (1997b). *World telecommunication development report 1996/97.* Geneva: Author.

Narain, R. (1996, June). A floorless exchange. *Asian Communications*, pp. 50–51.
Waller, E. (1997, June). All systems go. *Asia-Pacific Satellite*, pp. 40–42.

14

Participatory Politics and Sectoral Reform: Telecommunications Policy in the New South Africa

Robert B. Horwitz
University of California, San Diego

The April 1994 South African elections marked a monumental political transformation. After decades of minority rule based on racial domination and legalized violence, the White minority regime gave way to a democratically elected government of national unity led by the long-banned African National Congress (ANC).[1] Notwithstanding the immensity of the political change, the next and essential phase in the evolution of the "new South Africa"—economic development and redistribution—promises to be difficult and fraught with dangers. Restructuring the economy to facilitate long-term development and redistributing wealth and resources to the formerly dispossessed majority are both necessary. They may, however, be mutually exclusive. Redistribution has short-term political benefits but long-term economic risks. Restructuring is hoped to secure long-term economic benefits, but runs very real political risks, particularly for the liberation-identified ANC.

The new South Africa has been joyously welcomed back into the community of nations. This, too, has some ambiguous consequences. Rejoining the international community means the reassessment of domestic policies by a set of international

---

1. The basic agreement of the South African transition called for a legally mandated 5-year Government of National Unity regardless of the election outcome, with cabinet representation for all parties winning at least 5% of the vote and a share of executive power (an executive deputy presidency) to any party winning 20%. The ANC won 62.6%, the National Party won 20.4%, and the Inkatha Freedom Party won 10.5% of the 1994 vote. This gave three parties a corresponding share of cabinet portfolios (see Reynolds, 1994).

market standards. This, at bottom, is what "globalization" really means. To join the club—the international political economy—you have to consent to the club's rules. The risks of not joining the club are probably greater than the risks of joining, but joining means a reduced autonomy for national governments, entailing not simply a commitment to reduce protective tariffs and a move away from import substitution industrialization strategies, but also more generally a decline of Keynesian macroeconomic stabilization polices and state intervention into domestic economies. South Africa has been discovering these pressures in the context of World Trade Organization negotiations. Thus at the moment when the demands on the new South African state may be greatest, the resources are few, political maneuverability is restricted, and the range of possible reform is constrained by powerful macroeconomic forces.

One means of bridging economic restructuring with redistribution is to expand access to infrastructure services such as water, electricity, telephone, and transportation. This chapter examines the effort to transform one of those infrastructure sectors—the telecommunications sector. It tells the story of the reform of South African telecommunications from an economic sector organized to serve Whites and business, and dominated by an apartheid-structured state-owned enterprise, into something more democratic and efficient, capable of delivering the material goods to the Black majority. Indeed, telecommunications policy may be paradigmatic of the reform processes going on in contemporary South Africa. It gives people voice, symbolically and materially. Given the reregulatory discussions going on in telecommunications in many nations, this chapter explores what, if anything, is special about the South African situation. The story suggests that institutional reform, although both pushed and constrained by forces exogenous to a country, is still primarily a matter of political and economic forces endogenous to the country. And, although there are tensions between participatory and electoral structures, the South African example shows that good public policy can be made by stakeholders in open, deliberative, political processes.

SOUTH AFRICAN TELECOMMUNICATIONS AND PARASTATAL POLITICS: A BACKGROUND

The South African Posts & Telecommunications (SAPT), otherwise known as the Post Office, was a classic post, telephone, and telegraph (PTT) monopoly. As a state-owned monopoly or *parastatal*, the Post Office realized economies of scale and scope through vertical integration, ensured interconnection, and technical standardization. Its monopsony buying power assisted in the creation of a national equipment supply industry. Because it managed a complete system, the telephone monopoly could engage in the cross-subsidization of tariffs to expand service to particular constituencies. Like many other telecommunications regimes, the SAPT utilized value of service pricing and cost averaging schemes, among others, to effect cross-subsidies: typically, from business users to residential users, from long-distance (especially international) to local, from urban to rural. What was different about telecommunications in South Africa, of course, was apartheid. SAPT provided service primarily to Whites and to business. Telephone penetration per 100 Blacks was 2.4 in 1989 compared to a figure of 25 for Whites (Coopers & Lybrand, 1992).

In the late 1970s and early 1980s, however, the relative neglect of the non-White areas changed. The Post Office began to extend service to Indian and Coloured areas and, later, to some Black areas. This change was due to several factors, not the least of which was that the White telephone market was essentially saturated. At the broader level, the change also reflected a general, if desultory, set of government policies to retreat from some socioeconomic features of grand apartheid. These policies included making efforts to extend public services to non-Whites and to upgrade Black residential areas. Some of the most offensive features of apartheid, such as segregated entrances, elevators, waiting rooms, toilets, and benches in parks, had been gradually removed by the late 1970s. Perhaps most consequentially, the government moved to implement various labor reforms, including the scrapping of jobs reserved for Whites, bringing Black trade unions into the statutory industrial relations system, advancing the principle of equal payment for work of equal value, and abolishing segregation regulations under the laws relating to factories, shops, and offices (Giliomee & Schlemmer, 1990). This was the doing of the P.W. Botha *verligte* or reform wing of the National Party, a strategy that came to be known as *reform apartheid:* The ideological move of the reform strategy was to separate apartheid, now identified as a set of mechanisms that had outlived its functions, from White dominance, always the goal. The government's broad hope was to cultivate a non-White middle class whose political moderation would undermine opposition to continued White rule (Adam, 1971; Price, 1991).[2]

Contrary to the government's hopes, its reform strategy served to reawaken Black opposition. The government's reforms galvanized the widespread but splintered Black opposition, resulting in the formation of the United Democratic Front (UDF) in 1983. The UDF, organized primarily at the local level and embracing a deliberately vague nonracialism, engaged in community-based opposition to the government. This escalated into broad insurrection and precipitated the declara-

---

2. Three government-commissioned reports, produced by the usual cream of Afrikaner intelligentsia, emphasized that the time had come for the state to play a less dominant role in shaping the economy, that the private sector should play a much larger part in economic policymaking, and that the market should be left to operate as freely as possible. The severe labor and skills shortages South Africa had experienced for several years were laid directly at the door of apartheid policies. Indeed, two of the commissions seemed to have rejected White supremacy, arguing that the continuation of apartheid would undermine the continued success of the private enterprise system (Republic of South Africa, 1978, 1979a, 1979b).

tion of martial law (the state of emergency) in 1985 (see Lodge, Nasson, Mufson, Shubane, & Sithole, 1991; Marx, 1992).

Reform apartheid also necessitated an expansion of the state sector, accelerating a trend that had been in place for many years. Of particular notice was the growth of the parastatals. The percentage of the economy accounted for by the public corporations (not including the state business enterprises, such as the Post Office) had risen steadily through the years, and jumped rapidly between 1975 and 1985, from 4.1% of the gross domestic product (GDP) in 1975 to 7.7% of the GDP in 1985 (Nattrass, 1988). More important, perhaps, approximately two thirds of all capital available for investment was gobbled up by the public sector, and in particular by the three main infrastructure parastatals (the electricity parastatal Eskom, SAPT, and the South African Transport Services) in the decade 1971 to 1980 (Republic of South Africa, 1984). In telecommunications, expansion entailed large expenses for new local loops and a consequential move to digital switching. This necessitated extensive borrowing on international capital markets. The real value of that debt grew as the Rand plummeted consequent to the decline in gold prices and the confidence of the international financial community later in the decade.

As the National Party began its move away from grand apartheid, it also began moving toward classical market economics. Between the succesful historical project of Afrikaner economic uplift and a broader concern about apartheid's negative impact on the future of the economy, the reform (and dominant) wing of the National Party moved closer to the traditional economic concerns of business and English-speaking Whites in general. Here the example of the West, and the U.K. in particular, with its dramatic privatizations of British Telecom and British Gas in the Thatcher period, had real influence. For, notwithstanding the long contentious relationship between Afrikaans and English speakers, and between the National Party and the English Commonwealth, the U.K. has remained something of a model for White South Africa. Privatization was put forward as early as 1986 and was adopted as part of the government's long-term economic strategy in 1987 (Republic of South Africa, 1987).

However, parastatal reform in South Africa was far more complicated than the neoliberal economic policy it was presented to be. Privatizing the parastatals could accomplish several goals consonant with reform apartheid: it might address real operational and managerial problems in parastatals, bring money into government coffers through their sale, mollify the government's critics in the business community, *and* maintain White control over the parastatals through private shareholding. The political effect of privatization would be to take the parastatals out of the hands of the Black majority come a democratic dispensation. And, indeed, in early 1990, the newly legalized ANC announced that, although it was no longer wed to nationalization as a matter of general policy, state-owned enterprises that were privatized prior to political accommodation would be prime candidates for (re)nationalization (Battersby, 1990). This halted the privatization gambit. Commercialization, a status that had been viewed as merely a preparatory stage for privatization, then assumed importance as privatization's substitute. Accordingly, legislation separating posts from telecommunications and setting both set free from direct ministerial control was passed in October 1991 (Republic of South Africa, 1991). The separation created the telecommunications company Telkom, which, although it remains state owned, pays dividends and taxes, and is largely responsible for securing its own financing.

The Post Office Amendment Act did not, however, resolve the institutional and regulatory frameworks for the telecommunications sector. The central questions confronting any transformation of a PTT system—that is, which areas of service will remain monopolized and which will be competitive to what degree, and how should tariffs be adjusted—were left unaddressed by legislation. Moreover, questions of governance—that is, how and who would decide the central policy matters—were also not settled. After the 1991 act and creation of Telkom, telecommunications policy stalled.

THE POLITICS OF THE TRANSITION PERIOD: THE CIVIL SOCIETY THRUST

A central reason why movement in telecommunications policy languished was the failure of the sector to establish a stakeholder forum. Such forums sprouted from the politics of the South African transition, in which the National Party still held the reins of power after 1990 and continued to function as government, but had little legitimacy. The liberation movement championed the forums as a means to prevent the government from making decisions alone, particularly as constitutional negotiations were in motion and elections would presumably be held in the future. The forums represented the effort by excluded, largely Black groups to gain entry to policymaking arenas. Although local forums had been operating in some communities since the mid-1980s—and thus the pedigree for the grass-roots, consultative orientation of the forums lies in the township civic associations that grew during the UDF—the impetus for the formation of national forums was trade union politics. The Black labor umbrella organization, the Congress of South African Trade Unions (COSATU), believed that the National Party government had been unilaterally placing crucial areas of the economy outside the reach of political decision making in an attempt to limit the power of a future majority government. Following a general strike in November 1991 over the value-added tax (VAT), COSATU demanded a macroeconomic negotiation on social and economic issues, parallel to the political negotiations. The Business Roundtable, an embryonic business association anxious to rationalize macroeconomic policies and labor relations, was quick to support this move, as was Finance Minister Derek Keys. The National Economic Forum (NEF) was launched. Thereafter, an explosion of forums brought various constituencies together on all manner of issues at national, regional, and local levels to discuss matters such as housing, the VAT system, drought relief, and electricity. An estimated

230 forums grew in the period after February 1990 (Patel, 1993; Shubane, 1993).

The forums emerged in the nether-world of the period between the disintegration of the *ancien régime* and the transition to a new political dispensation. The township civic organizations and the stakeholder forums must be understood in the concrete South African political context of the resistance to White rule and later the transition to democracy and majority rule. The civics had functioned as loci for intense opposition to White rule and local self-help in the context of making the townships ungovernable during the liberation struggle. As part of organizing an opposition to White rule, many civics inaugurated participatory, consultative mechanisms for deciding on political strategies (Lodge et al. 1991; Shubane & Madiba, 1992). The later stakeholder forums adopted this ethos of participation and consultation. The legitimacy of the forums rested precisely in the fact that they took place outside the regular channels of the old government. At the same time, the government felt compelled to participate in the forums because any policy it might undertake risked being vetoed by the ANC alliance through strikes and street action if government proceeded without agreement from the forums. Although some of the forums, such as the NEF, were corporatist in membership and orientation, many others were broadly democratic in terms of representativeness, with specific participation from previously marginalized groups of civil society, the civic organizations in particular. The operative slogan of the forums—and in South African politics generally in this period—was "a culture of consultation and transparency."

Inasmuch as many forum participants were civil society groupings loosely orbiting around the liberation groups-cum-political parties (ANC, Pan Africanist Congress, Azanian Peoples Organization), the forums constituted a venue for them to accumulate information, expertise, and engage in policy debate in advance of their participation in formal electoral politics. The importance of the forums to the ANC was reflected in its 1994 Reconstruction and Development Programme (RDP) document—the ANC's macroeconomic vision for a postapartheid South Africa. The Congress explicitly stated that the RDP "must work with existing forums, such as the NEF, the National Electricity Forum and the National Housing Forum, and must develop a more coherent and representative system on the regional and sectoral basis" (ANC, 1994, p. 91). The idea was that democratization of the state was not restricted to universalizing the franchise; democracy was held to be incomplete unless civil society was *directly* brought into the policymaking process.

The consequences of no forum in the telecommunications sector

However, the forum process had not gelled in the telecommunications sector. This had serious repercussions. First, the absence of a consultative process in telecommunications led constitutional negotiators to annul a previous decision to deal with broadcasting and telecommunications policy together, and instead proceed with establishing a regulatory authority for broadcasting separately while putting telecommunications on hold. Second, the absence of a forum induced the government, led in this instance by an old-style National Party minister, to use telecommunications to its own strategic political advantage. Hoping to reap the political benefits of fostering new telecommunications services, the government moved to revamp the structure and governance of the sector unilaterally. The most consequential action in this regard was the decision to license two cellular telephone network operators in 1993. The cellular tenders became a significant political battleground in 1993 and 1994 when the ANC alliance contended that the licensing of cellular telephony outside the political negotiating forum represented a unilateral restructuring of the industry, a form of privatization through the back door (Fanaroff, 1993).

The politics of the cellular tenders became quite hot in September 1993, with COSATU threatening strikes and the ANC threatening to revoke the licenses when it came to power. ANC President Mandela met with State President de Klerk about the cellular license controversy, and ANC Secretary General Cyril Ramaphosa was quoted as saying that a future government would immediately review and perhaps cancel the cellular licenses if the government went ahead and issued them (*Citizen*, 1993; Makhanya, 1993). The ANC eventually backed down from its opposition largely because MTN and Vodacom, the cellular tender winners, agreed to support Black business by bringing in significant percentages of Black shareholding (*Business Day*, 1993). Black business' stake in MTN would become a centerpiece of the emerging movement for Black economic empowerment.

During this conflictual 1991 to 1994 period, both Telkom and the sector as a whole suffered. Telkom, in poor financial shape and saddled with a heavy debt burden, installed new lines at lower than a 5% increase rate per year, hardly a rate capable of rectifying apartheid inequities (Telkom, 1996). This is not to claim that the various stakeholder forums were necessarily successful in coming to consensuses or formulating concrete, implementable policies. In many instances they were not. However, the forums were essential in getting parties previously virtually unknown to each other to comprehend the necessities and complexities of policy under the new political dispensation. And their symbolic importance, in bringing non-Whites to policy arenas as equals, was vital. The forums also functioned materially—they furnished the ANC alliance an instrument with which to hem in the government's policy options during the transition period. The forums, in short, not only were essential in "doing politics" during the transition period, but also in laying a broad foundation for political dialogue and the basis for any policy decisions in a postapartheid South Africa.

THE 1994 ELECTION AND THE NEW POLICY INITIATIVE

Three events consequent to the 1994 elections began to move telecommunications policy in the direction of resolution: the establishment of a National Telecommunications Forum (NTF); the RDP set goals for telecommunications development; and the new (ANC) Minister for Posts, Telecommunications and Broadcasting in the Government of National Unity, Z. Pallo Jordan, put in motion a Green Paper/White Paper process to develop telecommunications policy.[3] The formation of the NTF finally engaged government, business, labor, user groups, and civic organizations in the consultative processes that stakeholders in other sectors had established years earlier. The RDP set a specific goal that telephones should be provided to all existing schools and health clinics within 2 years, hence providing formal impetus to the rapid expansion of the network. The Green/White Paper process, which would take place outside the old regime-identified Department of Posts and Telecommunications, tied the politics of telecommunications reform to the forum structures generally and to the consultative mechanisms of the NTF in particular.

Minister Jordan initiated a Green/White Paper process outside his ministry, as he did not trust the old guard. A respected ANC intellectual, Jordan had been sharply critical of the compromise that committed the ANC to a government of national unity and the retention of the existing military and civil service bureaucracies (ANC, 1992; Jordan, 1992; Slovo, 1992). Jordan believed such an agreement would leave the old regime apparatchiks intact to sabotage transformation. Accordingly, as minister, Jordan established the National Telecommunications Policy Project (NTPP), which convened a Technical Task Team of local knowledgeable people broadly representative of the main players in the sector, assisted at punctuated intervals by a small group of international "experts." The effort was led by a trusted political player, Willie Currie, who had earlier been a principal in the struggle to reform the broadcast sector (see Louw, 1993). The members of the team were selected by Minister Jordan in consultation with the NTF, and the NTF was to have various points of input into the process.

The Green Paper was written in the spirit of stakeholder consultation that had become the hallmark of the forums. Task team members elicited input from the stakeholders at a series of meetings. The Green Paper, an 80-page document published in four languages and placed on the Web, was written in a direct and relatively simple manner. It attempted to provide basic information, analysis, and the elucidation of trade-offs. Its mode was to pose a series of questions divided by themes into chapters: market structure, ownership, regulation, human resource development, Black economic empowerment, and the like. Parties were given about

---

3. The Green/White Paper model follows the British style of policymaking with a South African twist. British Green Papers indicate the intention of government. The South African Green Papers consciously do not. They describe the nature of a problem and ask a series of questions or identify a set of policy options. The White Paper answers those questions, decides on a set of options, and constructs a blueprint for legislation.

4 months to respond with written submissions (Ministry of Posts, Telecommunications and Broadcasting, 1995a).

The submissions in themselves were not surprising. What was interesting was how they accommodated the new politics of the new South Africa. Telkom, commercialized and now headed by a liberal technocrat, took on the mantle of speaking on behalf of those without service, claiming that only under traditional conditions of exclusivity and cross-subsidization could the network be expanded fairly and rapidly. Whatever liberalization might be undertaken must be gradual, several years in the future, and must not undermine the company's efforts at securing universal service. Labor, to a point, found itself strategically allied with Telkom. It denounced any consideration of privatization and lobbied vociferously on behalf of the maintenance of Telkom's monopoly. However, labor was also suspicious of Telkom, particularly of the internal reorganization efforts the company had put in motion earlier in the year and that promised to hive off "non-core" units from the parastatal. For labor, only a continued strong state presence would protect jobs and ensure that the state asset would be used to the benefit of the Black majority. Business, although internally divided according to position vis-à-vis Telkom, kept up a steady drumbeat as to the inoperability of old structures and the inevitability of competition. Many submissions, particularly those from large business users, lobbied for quick privatization and immediate competition (Ministry of Posts, Telecommunications and Broadcasting, 1995c; National Telecommunications Policy Project, 1995; Technical Task Team, 1995).

Minister Jordan invited key representative stakeholders (including business, labor, Black economic empowerment groups, relevant government departments, user groups, and civics) to a national colloquium in a country hotel for an intensive 3-day set of meetings to discuss how to proceed on the key policy questions. Various small working groups were constituted to try to arrive at consensus, which then reported back to plenary. There the process became partly stuck. Labor representatives could not abandon their mandated positions stipulating full state ownership of Telkom and no liberalization of the sector. Taking full note of labor's positions, the plenary went ahead and arrived at a bracketed consensus: Telkom should be granted a limited period of exclusivity (between 3 and 5 years) in basic switched telecommunications services. Various market segments should be liberalized in a phased fashion by an independent regulatory body. Telkom should be encouraged to take on a strategic equity partner to enable it to expand the network rapidly to reach the previously disadvantaged. There should be no binding contracts with local equipment suppliers, and there was a recognition that import tariffs would fall to General Agreement on Trade and Tariffs (GATT) levels or below. The ownership question was directed to the minister. All stakeholders agreed to the establishment of an independent regulator and, after a heated discussion, a separate Universal Service Agency to keep the universal service focus paramount (Ministry of Posts, Telecommunications and Broadcasting, 1995b).

A fear of reform being hijacked by White business lay behind the controversy over establishing a separate Universal Service Agency to operate alongside the regulator. The reality of political debate in the new South Africa is that, at least with regard to certain technical sectors like telecommunications, the apartheid legacy lives on. The Black participants in these discussions posed the basic sectoral policy objectives with eloquence and conviction, but when the discussions took a technical turn, whether with regard to technologies or economics or existing law or international practice, those with experience in and knowledge of the sector—overwhelmingly White—took over. Under the cordial surface of the discussions simmered an intense distrust. Although everyone professed to support the universal service aim, many Blacks suspected it would be abandoned by White business as the sector proceeded toward normalcy and business-as-usual. Black representatives, led by labor, insisted on the establishment of a Universal Service Agency for fear that the future regulator would be captured by White business interests. Symmetrically, business, worried about the politicization of regulation, lobbied strongly for the principle that the future regulatory bodies be staffed by knowledgeable and qualified people.

Bracketing any reference to the financing of network expansion and ownership (specifically with respect to privatization and levels of foreign direct investment), the NTPP Technical Task Team, along with reconvened internationals, wrote up the basic consensus of the Colloquium in a first draft of a White Paper. This would involve a fair amount of creative writing. Although the Colloquium resolved several broad policy items, it did not or could not determine the intricate mechanisms that would facilitate a period of exclusivity for Telkom followed by a phased liberalization. An Eminent Persons Group (EPG) was formed to make sure the team correctly embodied the Colloquium consensus, although in reality, the EPG, which consisted of representatives from Telkom, labor, the broadcast signal distributor, the Electronics Industries Federation, and Black business, converted the oversight discussions into one more negotiating forum.

Elements of the White Paper

The basic elements of the White Paper, the basis for draft legislation to go to a vote in Parliament, were as follows:

- Telkom is granted a limited period of exclusivity to provide most basic telecommunications services, thus allowing it to expand the network as rapidly as possible to facilitate universal access, first, and to move toward providing universal service.[4] The period of exclusivity has the additional goal of facilitating Telkom's preparation for eventual competition.
- Customer premises equipment (CPE) are completely deregulated.
- Resale of communications service by other private entities is permitted in Year 4.

- Telkom will rebalance its tariffs by Year 4.
- A Universal Service Fund will collect contributions from competitive segments to cross-subsidize areas where the infrastructure must be built and to customers who cannot afford regular tariffs. The preferred method of subsidy is one of targeted subsidies to end-users.
- National long distance will be opened to competition in Year 6.
- International services and local loops will be opened to competition in Year 7. Entry into these markets requires a license, and competition is understood as regulated competition.
- Telecommunications networks operated by other parastatal organizations (Eskom, the electricity parastatal, and Transnet, the transportation parastatal) are not permitted to compete with Telkom's service offerings. They are to "complement" Telkom's network.
- All this is to be overseen by a strong, independent regulatory body, the South African Telecommunications Regulatory Authority (SATRA). The Ministry of Posts, Telecommunications & Broadcasting will formulate general policy for the sector and assume the responsibility of acting as Telkom's shareholder.
- Interconnection mechanisms and charges will be a matter of contracts between private parties, with SATRA able to oversee and compel agreements and their terms where necessary.
- The self-provision of links to the network is permissible where and when Telkom cannot accommodate a service request with reasonable quality in reasonable time.
- A separate Universal Service Agency will maintain the focus on the universal service imperative and distribute Universal Service Fund subsidies.
- Strong efforts will be undertaken to expand the role of Black businesses in the sector and facilitate human resource development (Republic of South Africa, 1996b).

A key policy goal of the telecom reform process was to build consensus among stakeholders. This was achieved. The NTPP guided a very complex process of negotiations among virtually all of the telecom stakeholders, resulting in a set of compromises on market structure, ownership, commitment to universal service and independent regulation, and so on, and embodied in a technically detailed White Paper considered a triumph by most everyone. It represented a reasonable compromise between state control and international and domestic competitive

---

4. One of the Colloquium discussions concerned the difference between universal access and universal service. Because of the nature of Black settlement, particularly in rural areas and squatter settlements, most concluded that universal service—understood as telephone service to any and all dwellings—was unrealistic in the near future. Universal access meant the ability of people to access a public telephone according to reasonable time and distance factors (see National Telecommunications Policy Project, 1995).

pressures, and plotted a gradual, timed liberalization. The White Paper granted Telkom a period of exclusivity, but created mechanisms of contestability to push the parastatal to become more responsive and more efficient. The commitment to expand the network massively was designed to marry the imperatives of redistribution and development. Network expansion for universal access was also expected to mitigate the problem of potential labor retrenchments, inasmuch as such expansion will require considerable manpower.

As such, the actual policy prescriptions of the South African Telecommunications White Paper introduced no dramatically new initiatives (with the possible exception of the Universal Service Agency). The White Paper replicated in its own specifically local fashion many of the policies put in place by many other countries in both the developed and developing world: commercializing and separating operations from government, increasing private sector participation, shifting governmental responsibility from ownership and management to policymaking and regulation, and containing monopolies and developing competition. It committed the sector to a an immense roll-out of the network while also encouraging the development of sophisticated value-added services through phased and regulated liberalization of competitive entry—first at the margins, then at the core of the infrastructure. Its universal service regulatory mechanisms are in some ways less innovative than some recent experiments (see Tyler, Letwin, & Roe, 1995).

What was innovative in South Africa was the *process*. Unlike most other places in the world where the telecommunications regime has been or is being recast, telecommunications reform in South Africa was conducted within a democratizing context and was itself a democratic process of a unique participatory and deliberative kind. In many, if not most countries, telecommunications reform has been pushed by political and economic elites, whose ability to bring about policy transformations derives largely from the insulation of "reform" from normal political decision-making channels and distributive claims (see Petrazzini, 1995; Waterbury, 1992). In marked contrast, the South African telecommunications Green and White Paper process constructed a genuine public sphere in which all relevant parties had access and the ability to participate in ongoing discussions and negotiations in substantive, rather than merely symbolic, ways. Here was an instance of negotiations within civil society and between civil society and the state over the shape of a new political economy, where consensus building would have normative force for the participants.

To be sure, not all the participation was on an abstractly equal basis. Everybody essentially recognized that there were key players in the sector, and that those players had to arrive at some accommodation for policy to move forward. The key players in South Africa in the telecommunications sector in 1995, after the end of apartheid, an ANC electoral victory of almost 63%, and a government of national unity, were undoubtedly Telkom, labor, and to a lesser degree, the various telecommunications business interests. The boundaries of possible reform were constrained by the global dynamic of telecommunications liberalization and South Africa's participation in the World Trade Organization negotiations on basic telecommunications. Complicating the transformation process was the legacy of apartheid. Whites, or perhaps more accurately, White business, retained a dominance over technical knowledge. However, the consultations and negotiations between stakeholders were real and subject to considerable give and take. The process worked because it had the backing of the Minister—not as a means to rubber-stamp his own preconceived ideas, but as a genuinely open forum for stakeholder discussion and compromise.

The participatory and the electoral structures clearly exist in some tension. The danger of stakeholder politics (and corporatism in general) is that organized interests—particularly producer interests—dominate policymaking. Consumer interests have a much more difficult time gaining relevant access to corporatist forums, and it is nearly impossible to bring in the unorganized (see Cawson, 1986). To be sure, there was no "Union of Those Without Telephones" raising hell in the telecommunications reform process. Most of the relevant consumers were "user groups" of certain kinds of business customers. This is why some observers have put forward a serious case against the forums and, by inference, the stakeholder-driven Green and White Paper processes. Friedman and Reitzes (1995) argue that stakeholder politics allow interests and organizations that have not submitted themselves to the test of public election to exert as much power as, if not more than, representatives who have.

These are dangers, but they did not apply to the telecommunications reform process. The process was substantively open, the results structured but hardly predetermined. This may be what marks the essential difference between the South African Green and White papers and, for example, the U.S. public hearing process. Although the traditional public hearing policymaking process in the U.S. context is theoretically an open one, the mobilization of bias and the effective distribution of power means that, for the most part, most of the time, the real debate is between and among fractions of capital. In South Africa, in contrast, at least for the moment the balance of political and economic power is in flux, and the present political culture keeps questions of redistribution on the front burner. The attainment of universal service, rhetorically, if not actually, was the fundamental focus of the South African telecom reform process. The bill's consideration by electoral democratic structures—the minister, Cabinet, and Parliament—ensured that the general public interest, to the extent it may have been absent from the stakeholder compromise, was properly part of the reform process. The final parliamentary check presumably protects against the possibility of a narrow corporatism.

THE DENOUEMENT:
THE TRIUMPH OF ELECTORAL OVER PARTICIPATORY DEMOCRACY?

This is not the end of the story, however, and its denouement underscores not just the tension between participatory and electoral processes, but a broader set of divides within the ANC alliance between a social democratic, civic participation tendency and a tendency that marries neoliberalism to a "commandist" style of leadership (see Lodge, 1992; Louw, 1994). In March 1996, just weeks after the White Paper on Telecommunications had been officially published in the *Government Gazette*, Pallo Jordan was dismissed from the Cabinet. Jay Naidoo was given Jordan's portfolio. Although many were surprised and dismayed at Jordan's dismissal, there was some reason for optimism with regards to the telecom reform. Naidoo, as the former General Secretary of COSATU, was seen as perhaps the one person who could convince labor of the benefit of selling an equity stake in Telkom and hence resolve the bitter debate over ownership. At the time of his appointment, Naidoo assured the sector he would continue the reform process. In fact, as the White Paper was translated into a draft legislative bill, Naidoo and his advisers made important alterations. Minister Naidoo's rewrite of the draft telecom bill reamalgamated much of the authority the White Paper had given over to SATRA, the future independent regulatory body. The draft bill dropped all references to the period of exclusivity for Telkom and the liberalization timeline. The minister retained prerogative on nearly all matters of competition and liberalization (Ministry of Posts, Telecommunications and Broadcasting, 1996). It seemed Naidoo had returned to the old form of ad hoc ministerial control, although he rejected that interpretation. The ministry publicly insisted that the general features of the White Paper remained intact—their application would be through licensing rather than written into statute. But nothing in principle would prevent the minister from extending Telkom's exclusivity indefinitely. The ministry's final draft also excised any mention of SATRA's independence (Minister for Posts, Telecommunications and Broadcasting, 1996). Naidoo's apparent scuttling of the stakeholder consensus provoked a small firestorm among the business stakeholders and prompted NTPP chair Willie Currie to resign in protest (*Financial Mail*, 1996b).

In the end, Minister Naidoo's changes to the telecommunications White Paper were less radical than they appeared. Although his story changed several times, and he dismissed objections to the final draft bill as the complaints of "sectional vested interest groups," Naidoo did finally state publicly that Telkom would face full competition in 6 years and different segments of the market would be opened to competition along the way. The rules governing Telkom's period of exclusivity would be built into its license conditions (*Financial Mail*, 1996a; Hartley, 1996). In the final legislation, Parliament, over the objections of the ministry, restored a modicum of independence to SATRA (Republic of South Africa, 1996a). The draft Telkom licenses did limit the parastatal to a 5-year exclusivity in public switched service, with the proviso that it connect a set number of new lines and live up to service quality standards or face penalties (Republic of South Africa, 1997). With labor's qualified blessing, the government successfully negotiated the sale of 30% of Telkom to a consortium of SBC Communications and Telekom Malaysia.

The transformation of South African telecommunications was a success; the policy technically viable and for the most part politically legitimate. The process was threatened by the Minister's move away from consultation, but in the end he was forced to return to most of the main features of the White Paper. The participatory and electoral processes meshed fortuitously, if conflictually. Some commentators have implied that policymaking in postapartheid South Africa requires getting Blacks to accept an essentially choiceless economic reality that largely dashes their hopes and expectations (Adam, Slabbert, & Moodley, 1997). This cynically conceives democracy as a mode of generating consent behind prepackaged economic policies. It does not adequately capture the essence of the reform process. In the South African telecommunications reform process it seems nearly all parties learned the virtue of public reasonableness that has come to be a hallmark of contemporary liberal theory of citizenship (see Bohman, 1996; Gutmann, 1987). The stakeholder deliberative policy processes, exemplified by telecommunications reform, are about defining choices for a democratic political community within economic constraints.

ACKNOWLEDGMENTS

A longer version of this chapter appeared in *Media, Culture & Society, 19* (1997). The author wishes to acknowledge the Fulbright Foreign Scholarship Board, whose fellowship award permitted him to conduct research in South Africa.

REFERENCES

Adam, H. (1971). *Modernizing racial domination: South Africa's political dynamics*. Berkeley: University of California Press.

Adam, H., Slabbert, F., & Moodley, K. (1997). *Comrades in business: Post-liberation politics in South Africa*. Cape Town, South Africa: Tafelberg.

African National Congress. (1992, November 18). *Department of Information and Publicity, Negotiations: A strategic perspective (as adopted by the National Working Committee)*. Johannesburg: Author.

African National Congress. (1994). *The Reconstruction and Development Programme: A policy framework*. Johannesburg: Umanyano Publications.

Battersby, J. (1990, May 10). ANC tempers hard-line rhetoric on economic policy. *The Christian Science Monitor*, 3.

Bohman, J. (1996). *Public deliberation: Pluralism, complexity and democracy.*

Cambridge, MA: MIT Press.

Business Day. (1993, September 30). Compromise deal on cellular phones. Reported in *This Week in South Africa* (September 28–October 4).

Cawson, A. (1986). *Corporatism and political theory.* London: Blackwell.

Citizen. (1993, September 15). Government, ANC impasse on cellular phone controversy. Reported in *This Week in South Africa* (September 15–21).

Coopers & Lybrand. (1992). *Telecommunications sector strategy study for the Department of Posts and Telecommunications.* Pretoria, South Africa: Department of Posts & Telecommunications.

Fanaroff, B. (1993, August 31). National Secretary of Organising Department, National Union of Metalworkers South Africa (NUMSA), [Interview by author].

Financial Mail. (1996a, November 8). Jay's number dialed. Available: http://www.fm.co.za/

Financial Mail. (1996b, June 21). Naidoo's sleight-of-hand. Available: http://www.fm.co.za/

Friedman, S., & Reitzes, M. (1995). *Democratic selections? Civil society and development in South Africa's new democracy* (Development Paper 75). Halfway House, South Africa: Development Bank of Southern Africa.

Giliomee, H., & Schlemmer, L. (1990). *From apartheid to nation-building.* Cape Town, South Africa: Oxford University Press.

Gutmann, A. (1987). *Democratic education.* Princeton, NJ: Princeton University Press.

Hartley, W. (1996, October 8). Naidoo tables legislation "vital" to securing Telkom equity partner. *Business Day,* 1.

Jordan, P. (1992). Strategic debate in the ANC: A response to Joe Slovo. *The African Communist, 131,* 7–15.

Lodge, T. (1992). The African National Congress in the 1990s. In G. Moss & I. Obery (Eds.), *South African Review 6: From "Red Friday" to Codesa* (pp. 44–78). Johannesburg: Ravan Press.

Lodge, T., Nasson, B., Mufson, S., Shubane, K., & Sithole, N. (1991). *All, here, and now: Black politics in South Africa in the 1980s.* Cape Town, South Africa: David Philip.

Louw, P. E. (Ed.). (1993). *South African media policy: Debates of the 1990s.* Bellville, South Africa: Anthropos.

Louw, P. E. (1994). Shifting patterns of political discourse in the new South Africa. *Critical Studies in Mass Communication, 11,* 22–53.

Makhanya, M. (1993, September 17–23). Playing broken telephones. *The Weekly Mail & Guardian,* 18.

Marx, A. (1992). *Lessons of struggle: South African internal opposition, 1960–1990.* New York: Oxford University Press.

Minister for Posts, Telecommunications and Broadcasting. (1996). *Telecommunications bill* (as introduced), 15th draft. Dated July 12, 1996. Pretoria, South Africa: Government Printer.

Ministry of Posts, Telecommunications and Broadcasting. (1995a). *A Green Paper for public discussion: Telecommunications policy.* Pretoria, South Africa: Government Printer.

Ministry of Posts, Telecommunications and Broadcasting. (1995b). *Report on the proceedings of the National Colloquium on Telecommunications Policy* (M. Andersson, Compiler). Johannesburg: Author.

Ministry of Posts, Telecommunications and Broadcasting. (1995c). *Statistical analysis of responses to the Green Paper on Telecommunications Policy 1995—Interim report* (A. Bacchialoni & A. Wills, Eds.). Johannesburg: BMI TechKnowledge.

Ministry of Posts, Telecommunications and Broadcasting. (1996, June 7). *Fourteenth draft of Telecommunications Bill.* Pretoria, South Africa: Author.

National Telecommunications Policy Project. (1995). Submissions to Green Paper on Telecommunications Policy. Johannesburg: Various.

Nattrass, J. (1988). *The South African economy: Its growth and change.* Cape Town, South Africa: Oxford University Press.

Patel, E. (1993). *Engine of development? South Africa's National Economic Forum.* Johannesburg: Juta Press.

Petrazzini, B. A. (1995). *The political economy of telecommunications reform in developing countries: Privatization and liberalization in comparative perspective.* Westport, CT: Praeger.

Price, R. M. (1991). *The apartheid state in crisis: Political transformation in South Africa, 1975–90.* New York: Oxford University Press.

Republic of South Africa. (1978). *Interim report of the commission of inquiry into the monetary system and monetary policy in South Africa: Exchange rates in South Africa* [also known as the de Kock Report, after the commission chairman]. Pretoria, South Africa: Government Printer.

Republic of South Africa. (1979a). *Report of the commision of inquiry into legislation affecting the utilisation of manpower* [also known as the Riekert Report, after the commission chairman]. Pretoria, South Africa: Government Printer.

Republic of South Africa. (1979b). *Departments of Labour and of Mines, Report of the commission of inquiry into labour legislation* (Part 1) [also known as the Wiehahn Report, after the commission chairman]. Pretoria, South Africa: Government Printer.

Republic of South Africa. (1984). *Report of the commission of inquiry into the supply of electricity in the Republic of South Africa.* Pretoria, South Africa: Government Printer.

Republic of South Africa. (1987). *White paper on privatization and deregulation in the Republic of South Africa.* Pretoria, South Africa: Government Printer.

Republic of South Africa. (1991, June 19). *Government Gazette, Act No. 85 of 1991. Post Office amendment act.* Pretoria, South Africa: Government Printer.

Republic of South Africa. (1996a, November 12). *Government Gazette, Act No. 103 of 1996. Telecommunications act*. Pretoria, South Africa: Government Printer.

Republic of South Africa. (1996b, March 13). *Government Gazette, Notice No. 291 of 1996. The White Paper on telecommunications policy*. Pretoria, South Africa: Government Printer.

Republic of South Africa. (1997, May 7). *Government Gazette, Notice No. 768 of 1997. Licence issued to Telkom SA*. Pretoria, South Africa: Government Printer.

Reynolds, A. (Ed.). (1994). *Election '94 South Africa: The campaigns, results and future prospects*. Cape Town, South Africa: David Philip.

Shubane, K. (1993). *Tomorrow's foundations? Forums as the second level of a negotiated transition in South Africa*. Johannesburg: Centre for Policy Studies.

Shubane, K., & Madiba, P. (1992). *The struggle continues? Civic associations in the transition*. Johannesburg: Centre for Policy Studies.

Slovo, J. (1992). Negotiations: What room for compromise? *The African Communist, 130*, 36–40.

Technical Task Team, Ministry of Posts, Telecommunications and Broadcasting. (1995). *Interim narrative report on the responses to the Green Paper on Telecommunications Policy*. Johannesburg: Author.

Telkom. (1996). *Annual report*. Pretoria, South Africa: Author.

Tyler, M., Letwin, W., & Roe, C. (1995). Universal service and innovation in telecommunication services. *Telecommunications Policy, 19*, 3

Waterbury, J. (1992). The heart of the matter? Public enterprise and the adjustment process. In S. Haggard & R. R. Kaufman (Eds.), *The politics of economic adjustment: International constraints, distributive conflicts, and the state* (pp. 182–217). Princeton, NJ: Princeton University Press.

15

Telecom Competition in Canada and the United States: The Tortoise and the Hare

Willie Grieve
Saskatoon, Saskatchewan

Stanford L. Levin
Southern Illinois University at Edwardsville

The United States began the process of introducing competition into local telecommunications before Canada, and there are important differences in the approaches chosen by the two countries. The U.S. approach is based on public utility concepts, whereas the Canadian approach is based on antitrust and economic principles. This chapter shows that, as a result, Canada is likely to have widespread facilities-based competition for local telecommunications services before the United States.

INTRODUCTION

In the United States, a number of states have been introducing competition into local telephone service for several years. The U.S. Congress, with great fanfare, passed the Telecommunications Act of 1996, with the stated purpose of promoting competition and reducing regulation in order to achieve lower prices, higher quality services, and rapid deployment of new technologies for Americans.[1] It set out, in considerable detail, the way in which the local markets were to become com-

---

1. *See* Joint Statement of Managers, S. Conf. Rep. No. 104-230, 104th Cong., 2d Sess. Preamble (1996) (Joint Explanatory Statement); *see also* 47 U.S.C. § 706 (a) (encouraging the deployment of advanced telecommunication capability to all Americans).

petitive, and the Federal Communications Commission (FCC) implemented the provisions of the Act with a series of even more complex and detailed orders.[2]

At the time the Telecommunications Act of 1996 was passed, the Canadian Radio Television and Telecommunications Commission (CRTC) was engaged in a regulatory proceeding to establish rules for the introduction of local competition in Canada. The CRTC released its decision on May 1, 1997 (CRTC Local Competition Decision 97-8, 1997).

The models used in the United States and in Canada are quite different, however. This chapter argues that the United States began the race to a fully competitive telecommunications market quickly and took a big lead over Canada. The Canadians, however, by working slowly and methodically through competition principles, will cross the finish line first while the United States languishes in the shade tree of the Telecommunications Act of 1996. This is a shade tree that distorts the notion of competitive telecommunications to such an extent that it may actually entrench monopoly and market power in the local networks of the incumbent local carriers.

Although many of the same parties and witnesses appeared in the United States and in Canada, by the time the arguments were made in Canada they had become more refined. In particular, the consequences of network component unbundling and mandated resale were better understood, and this is reflected in the CRTC's decision.

This chapter first offers some comments about telecommunications competition and regulatory policy. It then illustrates the connection between the development of competitive markets and regulatory policy by analyzing "competitive" policies in the United States and Canada.

THE NATURE OF COMPETITION

In telecommunications, it is important to distinguish between facilities-based competition and services competition arising through the resale of the incumbent's facilities and services. One can certainly create the illusion of full competition by quickly creating regulatory incentives for resale and pointing to all the "competitors" in the market. However, a truly competitive telecommunications market will

---

2. FCC, Implementation of the Local Competition Provisions of the Telecommunications Act of 1996 and Interconnection between Local Exchange Carriers and Commercial Mobile Radio Service Providers, CC Docket No. 96-98 and CC Docket No. 96-185, August 8, 1996, First Report and Order, FCC 96-325; FCC, Federal-State Joint Board on Universal Service, CC Docket No. 96-45, May 7, 1997, Report and Order, FCC 97-157; and FCC, Access Charge Reform, Price Cap Performance Review for Local Exchange Carriers, Transport Rate Structure and Pricing, and End User Common Line Charges, CC Docket No. 96-262, CC Docket No. 94-1, CC Docket No. 91-213, and CC Docket No. 95-72, May 7, 1997, First Report and Order, FCC 97-158.

include competition in the provision of infrastructure or the underlying network, or, in other words, facilities-based competition, insofar as it is economically possible. If competition is only effected through the resale of the incumbent's underlying network elements and services, then all providers will become dependent on the incumbent's infrastructure, which will continue to be regulated. Although regulators may claim that they are establishing prices for facilities and services purchased by competitors at "competitive" levels, those prices are simply a theoretical guess as to what prices in a competitive market might be. Similarly, the quality of service and all other terms and conditions of service are simply a regulator's guess of what a competitive market would produce. If competition for infrastructure is possible, more benefits will be delivered to consumers.

COMPETITION AND REGULATORY POLICY

Competition is fundamentally at odds with economic regulation (i.e., the regulation of prices, terms and conditions of service, obligations to provide service, and quality of service monitoring). If a market is regulated, it cannot be truly competitive. This leads to two conclusions. First, to achieve the objective of competitive telecommunications markets, markets must be deregulated when at all possible. Second, if significant regulation persists, markets may not look competitive and, therefore, regulators will not consider deregulation to be in the public interest (however defined). This circularity, if it is not broken by deregulation when the market would, in fact, be competitive in the absence of regulation, may prevent competition and its attendant benefits from actually arising.

Policies designed to introduce competition into formerly monopoly markets should be based on the economic and legal principles of competition rather than the policies of public utility regulation. Public utility status and the accompanying regulation arose because of the simultaneous presence of two conditions, natural monopoly and the essentiality or importance of the service to the public. There can be no question that telecommunications remains an essential or important service. The question of whether the local network remains a natural monopoly, however, is still being debated in the United States, in spite of facilities based entry. That debate, however, has all but ended in Canada. In its Local Competition Decision, the CRTC, based on evidence of actual and planned competitive facilities, found that there are only three essential (or natural monopoly supplied) facilities: telephone numbers (NXX[3] office codes that will be administered by an independent third party), directory listings (name, address, and telephone number, which can

---

3. NXX is the standard telephone industry abbreviation for a telephone number prefix (the first three digits of a seven digit phone number) in which the first digit of the prefix is restricted to certain numbers (0 and 1 are not allowed, for example) but the second and third digist of the prefix may be any number.

only be provided by the carrier serving the customer), and local loops in certain high-cost areas (from the customer side of the main distribution frame to the customer's premises). What this means is that the local exchange carriers in Canada have been found by the CRTC not to be natural monopolists except to the extent they provide local loops in high-cost areas (and new wireless technology may end this natural monopoly as well[4]). Infrastructure competition should therefore be feasible except, at least at this time, in high-cost rural areas.

Because the natural monopoly condition that gave rise to public utility status for incumbent local exchange carriers has largely disappeared, so too must the public utility-style regulation that was developed to deal with that condition. Because competition can now arise in local telecommunications, it is competition law and economic principles that must guide the development of competition.

This section of the chapter considers two key aspects of local telecommunications policy: (a) the degree of mandated unbundling and the pricing of unbundled facilities, and (b) mandated resale of local service and any wholesale discounts. First, however, it is helpful to reiterate the distinction between interconnection and unbundling.

Interconnection and Unbundling

Interconnection is the connection of all local telephone networks so that the customers of any carrier are able to call the customers of any other carrier. Both Canada and the United States recognize that interconnection is a requirement to have competitive local markets. *Interconnection*, as the expression is used in this chapter, is simply the provision by each carrier of the call termination function at its end office local switch. The originating carrier is responsible for arranging for the carriage of a call made by one of its own customers to the called customer, whether that called customer is its own or is served by another carrier. Where the customer is served by a second carrier, it is only that second carrier that can terminate that call to a specific telephone number. It is at that point, the terminating point on the switch serving the called customer, that the interconnection actually takes place.

Unbundling, on the other hand, although it requires the connection of facilities to accomplish, is the provision of piece parts of one carrier's network to another carrier in order to allow the second carrier to construct its network using piece parts of the first carrier's network. Once the second network is constructed in this fashion, it must also be interconnected with the first carrier's network so that calls can be exchanged between customers on the separate networks.

---

4. A study by Hatfield Associates, Inc. (1994), *The Cost of Basic Universal Service*, showed that the incremental cost of providing wireless service in regions with less than 10 access lines per square kilometer was up to 40% less expensive than wireline service.

For any particular call to a particular telephone, only one carrier, the carrier serving the called customer, can terminate the call. The call termination is interconnection, and the lease of any of the facilities along the way from the originating customer to the end office switch serving the called customer is unbundling of the transport and intermediate switching. So, too, would be the provision of any other facilities and services (such as databases and other facilities and services required for the network to function) that one carrier leases from another.

It is important to make the distinction between interconnection and unbundling because two completely different policies underlie the decisions to mandate interconnection and to mandate unbundling. In the case of interconnection, the policy rationale is simply to ensure that all customers, regardless of the local carrier they choose, must be able to call and receive calls from all other customers. The rationale for unbundling finds its roots in the competition law principle of essential facilities, not the public policy principle that justifies mandated interconnection.[5]

This distinction was made and applied by the CRTC in Canada and is also made, although not as cleanly, by the Telecommunications Act of 1996. The problem is that the Telecommunications Act of 1996 does not employ the expression "interconnection" consistently. For example, in section 251, the Act imposes a general interconnection requirement on all carriers. It also requires all local exchange carriers (LECs) to provide interconnection at any technically feasible point in the network for the transport and termination of traffic. The FCC concluded that the first type of interconnection simply referred to the physical connection between facilities of carriers. The second type of interconnection, for transport and termination, was recognized by the FCC for what it really is, unbundling. The FCC was confronted by this reality when it discovered that unbundled transport and transport for call termination were one and the same thing and, therefore, had to be provided at the same price by the incumbent local exchange carriers (ILECs).[6] Therefore, even though the language employed by Congress was unclear, the FCC required each carrier to provide the call termination function to each other carrier at the local switch serving the called customer, at the same time recognizing that transport (and, incidentally, intermediate switching) is really an unbundled service that could be competitively provided.

Nevertheless, the confusion persists because Congress employed the "interconnection at any technically feasible point in the network" expression to describe the interconnection duty imposed on the ILECs. In order to make sense of this expression, the FCC listed a number of technically feasible points of interconnection that

---

5. It is possible to use the essential facilities doctrine to justify the requirement that all carriers provide the call termination function to all other carriers. However, this idea has problems of its own. These are discussed in Grieve and Levin (1996).

6. FCC, Implementation of the Local Competition Provisions of the Telecommunications Act of 1996 and Interconnection between Local Exchange Carriers and Commercial Mobile Radio Service Providers, CC Docket No. 96-98 and CC Docket No. 96-185, August 8, 1996, First Report and Order, FCC 96-325, Section XI. A. 2. c. (1).

are, in reality, the provision of unbundled transport and/or switching. The end of-fice local switch, the last place on the network to which the call destined to the cus-tomer can be delivered by the first carrier to the second, is actually the only interconnection point possible.

This can be illustrated with the figures presented here. In Figure 15.1, if LEC-B, the entrant, connects to one of LEC-A's, the incumbent, unbundled loops on the customer side of the main distribution frame (point X, a point that the FCC held was a technically feasible interconnection point), no "interconnection" has taken place because the customer served by that loop, now a LEC-B customer, cannot call LEC-A's customers unless LEC-B also interconnects with LEC-A in order to exchange calls (point Y). Point X is a connection to gain access to an unbundled loop, whereas point Y is an interconnection so that LEC-B (the purchaser of the unbundled loop) can terminate calls from its own customer, now served through an unbundled loop, to other customers still served by LEC-A.

FIGURE 15.1
Unbundled loops and interconnection.

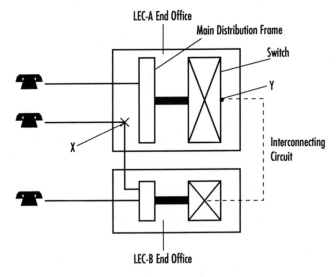

In Figure 15.2, the two LECs operate separate networks and must interconnect for the exchange of calls. Under the U.S. Telecommunications Act of 1996, LEC-B is entitled to interconnection at any technically feasible point. It chooses the LEC-B side of the tandem switch (Z). Therefore, LEC-A must provide to LEC-B tandem switching, transport to the local switch, and the call termination function at the local switch. However, under the terms of the U.S. Act, LEC-B has another option. It may choose to purchase unbundled tandem switching and unbundled transport to reach the call termination point (CT). In order to do so, it would connect at precisely the same point (Z) and receive unbundled tandem switching and unbundled transport to the local switch (CT) for call termination. Therefore, interconnection at any technically feasible point is nothing more than another way of mandating that everything be unbundled.

FIGURE 15.2
Interconnection at any technically feasible point and unbundling.

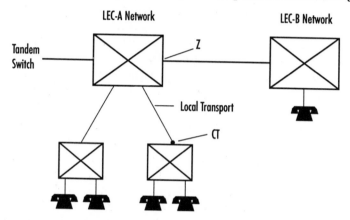

There can be no competition unless interconnection, the provision of the call termination function by all carriers to all other carriers, is present. In both Canada and the United States, interconnection for call termination is mandated. Where the two countries differ is in the amount of unbundling that has been mandated, and this difference is one of the reasons the Canadian approach will produce full competition (facilities-based competition) more quickly.

Mandated Unbundling and the Price of Unbundled Facilities

With the prospect of the introduction of local competition, potential competitors typically argue that they need to be able to purchase any portion of the incumbent carrier's local network so that they can establish their own service. The first competition question is how much mandatory[7] unbundling the incumbent carrier

---

7. The mandatory nature of this unbundling is key. Incumbent carriers may choose to make available portions of their facilities or services on a voluntary basis based on their analysis of the business opportunity and competitive conditions in the facilities market. Such voluntary unbundling is not at issue. Only when regulators mandate that the incumbent make available certain facilities or services to competitors does the economic and public policy question arise.

should be forced to undertake. This question has been debated and answered by antitrust or competition law. The answer is clear. Only essential facilities should be subject to mandatory unbundling. If a facility is essential, a potential competitor cannot compete without the use of such a facility. Further, the market for that facility cannot be competitive. Therefore, the terms on which the essential facility is provided to competitors, including its price, must be regulated.

Essential facilities can be defined as follows: "An essential facility is a monopoly-supplied facility, function, process, or service which competitors require as an input in order to provide telecommunications services and which competitors cannot economically or technically duplicate" (Grieve & Levin, 1995). In Canada, the CRTC adopted a definition of essential facilities that for all practical purposes mirrors this definition (CRTC Decision 97-8, 1997).

Problems arise, however, when mandated unbundling goes beyond essential facilities. In that case, incumbent local carriers are forced to make available to competitors facilities that could be provided by alternative suppliers or that could be economically self-supplied by the would-be competitor. Economically supplied means that the cost incurred by the alternative supplier entering the market at minimum efficient scale would be no higher than the cost incurred by the incumbent to supply the facility.[8] In other words, to justify mandatory unbundling, the facility or service in question must exhibit the properties of natural monopoly. So, where public utility regulation focused on the natural monopoly provision of essential services to end users, the regulation of essential facilities focuses on the provision of natural monopoly-supplied facilities to competitors so that competition can arise for the remainder of the services and facilities necessary to provide service.

If incumbents are forced to make nonessential facilities available to their competitors, the incentives are to curtail research and development and innovation in the production of the facilities (or alternatives to those facilities) that they are mandated to share, as the incumbent would capture no gains relative to its competitors. Indeed, Areeda (1989) argued that it is the recognition of the innovation-chilling effects of such a policy that is one of the fundamental rationales for the development of the essential facilities doctrine and the need to limit mandated sharing (unbundling) to essential facilities properly defined.

The decision to mandate the unbundling of nonessential facilities is, itself, anticompetitive, but the regulated price at which these unbundled facilities must be made available to competitors also has important competitive consequences. Telecommunications carriers have a cost structure that includes shared and common costs due to the multiproduct and network nature of the services they provide. If a

---

8. If minimum efficient scale in a market is greater than 50% of the market, the would-be competitor would not be able to enter and compete unless it substantially supplanted the incumbent. In this case, the incumbent would not be operating at minimum efficient scale and competition would not be self-sustaining.

carrier sold every service at incremental cost, the carrier would not recover its total costs as some or all of the shared and common costs would not be recovered.

With this cost structure, unbundled facilities should not be sold at incremental cost but, rather, at incremental cost plus a markup to cover some shared and common costs. There are two reasons for this. First, such pricing will be more efficient because it will approximate the price that would be charged in a competitive market. Second, such pricing will not stifle the emergence of competition or create the appearance of an essential facility when one does not or no longer exists. If an unbundled facility were priced only at incremental cost, it would always be more attractive for an equally efficient competitor to purchase the unbundled facility from the incumbent at incremental cost, because any other service provider's price would necessarily have to be higher in order to cover some portion of that provider's shared and common costs.

Both the mandatory unbundling of nonessential facilities and too-low pricing for these unbundled facilities are, then, anticompetitive, as such policies discourage the entry of competitors into the market that might offer service through their own facilities in competition with the incumbent carrier. Widespread mandatory unbundling actually entrenches the incumbent carrier's monopoly or market power in the network and perpetuates regulation instead of permitting competition to arise.

Resale of Local Exchange Service

Potential local exchange competitors have argued forcefully that they should be able to buy local exchange service from the incumbent and then resell it to precisely those same customers. Such a policy, in effect, mandates the creation of a wholesale local market unlike anything that one observes in nonregulated competitive markets.

Whereas firms in competitive markets have the option of deciding whether to allow resale of their products or services, telecommunications firms have generally been required by regulation to permit resale of their services. Mandating resale, however, is inconsistent with competitive markets, in which suppliers would determine, based on an analysis of the business case, whether or not they wanted to allow the resale of their services. In this context, mandating resale has much the same consequences as mandating the unbundling of nonessential facilities: there is little incentive to invest in innovation and research and development if all new innovations must be made available to one's competitors.

The rationale for mandated resale has never been clear. It seems to rely on the assumption that this somehow hastens the achievement of competition in the market. Although it does mandate some version of competition for the retailing function, it may slow down or preclude facilities-based competition. Of course, if facilities-based competition is not possible due to the presence of natural monopoly, efficient resale is certainly preferable to the alternative of monopoly. One might argue that most of the value added from competition is currently in services built on top of a commoditized local infrastructure, although there is no evidence that this is true. To the extent that there is any additional value added to be gained from the underlying infrastructure, in any case it is certainly far less likely to be realized under conditions of regulated monopoly than under conditions of competition. This is a second best policy when the best policy is available.

Potential entrants have also argued that they need a certain size discount off of the retail rate so that they can profitably resell local service, although why they should be able to do this at a profit has never been made explicit. Entrants may want to provide resold local service bundled with their long distance, cable television, or wireless service, but that does not mean that stand-alone resale should necessarily be a profitable business. Furthermore, there is no competitive justification for mandated discounts. If mandatory wholesale discounts exceed the actual avoided costs when wholesale rather than retail service is provided (adjusted for any additional wholesale costs that might be incurred), competitive facilities-based carriers may be foreclosed from entering the market, except in certain dense areas where costs are significantly below average.

Competitors often point to the resale experience in the long distance market, but the comparison is completely inapt. In the long distance market, resale off the regulated tariffed rates was mandated, but there were no mandated wholesale discounts. Resellers have been free to buy high-volume toll services available to high-volume retail customers and resell them to groups of smaller customers that the reseller has aggregated. There has never been a policy that required the incumbents to provide discounts off existing retail tariffs.

When an incumbent is mandated to provide a percentage discount off of a retail rate and to provide that discounted service to a competitor, there is no incentive for the incumbent to reduce its costs of providing service. If the costs go up, regulation will permit a rate increase for the retail customer and the "wholesale" customer, but if the costs of the incumbent go down, the incumbent cannot reduce its retail rate without also reducing the rate charged to its competitor. The competitor, in effect, has a regulatorily guaranteed margin. This is not competition, it is regulation.

Mandatory resale of local exchange service and mandated discounts can give the illusion of competition by making it easy for many companies to resell the underlying service of the incumbent, facilities-based carrier, but they may hinder, rather than help, the development of facilities-based competition.

U.S. AND CANADIAN REGULATORY POLICY AND ITS IMPLICATIONS

Public Utility Regulation and Antitrust

The major difference between the U.S. and Canadian approaches to these two major local competition issues, mandatory unbundling and resale, can be traced to the difference between public utility regulation and antitrust or competition law. In the United States, the Telecommunications Act of 1996 and the FCC's orders implementing the framework for local competition (see footnote 2) have as their foundation public utility concepts, even though, as we have argued elsewhere, public utility concepts should not be employed for the purpose of developing competition policy (Grieve & Levin, 1996). That the U.S. Act finds its intellectual underpinnings in public utility regulation is evident by making a few specific observations. The U.S. Act imposes additional requirements on incumbents. Those requirements are not explicitly based on a market power analysis; they are based on incumbency alone, and it is the incumbent who is impressed with an obligation to serve competitors through mandated unbundling of anything competitors might demand and through mandated resale discounts. It is the incumbents, also, that are impressed with the obligation to provide "interconnection at any technically feasible point in the network," which, as shown previously, amounts to another way of expressing the rule that everything is to be unbundled.

One could argue that although the Act does not specifically mention the concept of market power as a justification for the imposition of expanded public utility obligations on incumbents, the Act simply assumes market power. This is unlikely, however, because the Act applies public utility regulation tests to the analysis of when there might be sufficient competition to justify forbearance. For example, the regulator is to forbear where there is sufficient competition to ensure that rates are "just and reasonable" and not "unjustly discriminatory or unduly preferential." Those are public utility, not competition, tests. Because it is unlikely that competition will ever produce the same outcomes that regulators of public utilities consider to be just and reasonable and not unjustly discriminatory, it is unlikely that regulators, applying their own test, will ever find there is sufficient competition to permit forbearance. The wrong test is specified in the Act, and the wrong people are called on to make the determination. Congress chose competition and enacted regulation.

In Canada, the recent CRTC decision implementing local competition (CRTC Local Competition Decision 97-8, 1997) does not make the same mistake that the U.S. Congress and the FCC made. The Canadian decision is based on economics and competition law concepts. It recognizes the importance of developing rules that allow facilities-based competition to emerge and uses competition law concepts to do so. There is no expanded obligation to serve competitors. Significantly, also, the Canadian Telecommunications Act (enacted in 1993) does not employ public utility language in its forbearance section. Instead of requiring forbearance where there is sufficient competition to ensure that rates are just and reasonable and not unjustly discriminatory, the Canadian Act deliberately requires forbearance where there "is or will be competition sufficient to protect users" (CRTC Local Competition Decision 97-8, 1997). In Canada, the Commission must ask whether there is or will be competition sufficient to protect users as they would be protected by market forces in competitive markets, not by regulatory forces in regulated markets.

Unbundling

The differences in the fundamental approaches to the introduction of local competition in Canada and the United States are also evident in specific policies. The decision by Congress to employ traditional public utility-style regulation to introduce competition naturally leads to policies of widespread unbundling and mandated resale. This is so because the remedies and tools built into that style of regulation have as their underlying assumption the view of the network as a natural monopoly. Thus, in the United States, the Telecommunications Act of 1996 requires extensive unbundling of the incumbent's network (Telecommunications Act of 1996, Section 34(2)), and this unbundling goes well beyond what is required by the essential facilities doctrine. In general, incumbent local carriers are to make available to competitors any part of their network that can technically be offered. This includes loops, switching, transport, signaling, and probably subloop elements if they can technically be made available.

In the United States, the anti-competitive consequences of this extensive unbundling have been made worse by the pricing requirements. The Telecommunications Act of 1996 says that these elements are to be made available at incremental cost plus a markup to cover shared and common costs (Telecommunications Act of 1996, Section 251(c)(3)), although the recovery of a portion of shared and common costs may not be guaranteed. The FCC, in its local competition order, implements this portion of the Telecommunications Act of 1996 with a total element long run incremental cost (TELRIC) standard and a promise to recover a portion of the forward-looking shared and common costs (Telecommunications Act of 1996, Section 252(d)(1)). Because TELRICs are hypothetical, rather than any company's actual costs, and because of the actual methodology, they tend to set prices that are below the incumbent company's actual (either in historical or forward-looking terms) costs of providing the network elements. Similarly, the forward-looking nature of the shared and common costs means that these, too, may be understated when compared to actual costs.

There are a number of observations that lead to this conclusion. First, and most obviously, TELRIC costs can, by definition, never be greater than an incumbent company's actual costs. Second, TELRIC costs are likely to be below a company's actual long-run incremental costs because they are based on the least cost technology that would be available if the current network (adopting the same switch locations) were to be rebuilt from the ground up today. Of course, that will never happen, so to the extent that older and more expensive technology (which may have been the least cost technology at the time it was deployed) is still employed in the network, TELRIC costs must be lower than actual historical costs.

Third, TELRIC costs may be lower than actual incremental costs (those that would actually be incurred to provision facilities for an increment of demand) because there are many cases where the least cost solution to provisioning more capacity on an existing network is not to abandon the old technology and replace it from the ground up with new technology, but rather to provision for the increment of demand using more of the old technology. The point at which a feeder system is converted from copper to fiber is a good example of such a costing problem. It may very well be cheaper to have a full fiber feeder system operating on one route, but at the time the existing feeder system was built, fiber may not have been the least cost solution, and as the capacity on the route grew, no single actual increment of demand for feeder capacity could justify tearing up the copper and installing fiber. Switching provides a similar example. Why would a carrier discard an entire switch to meet an increment of demand when it could very inexpensively add another module to the existing switch? Finally, no communications carrier (incumbent or new entrant) will build its network instantaneously and rebuild it from the ground up every time a new technology is introduced. Therefore, to the extent that the cost of the least cost solution declines over time, any carrier that has already installed facilities will find that its average costs are higher than incremental costs. As a result, it cannot price simply to recover what its costs would be if it were to rebuild every day with the least cost technology of that day.

For all of these reasons, TELRIC costs can never be above an incumbent's actual costs and are likely to be below those costs. From any incumbent's point of view, then, prices that are based on TELRICs may be below cost. More worrisome, it may turn out that TELRICs are below the actual costs of many or all of the potential entrants. Even though the U.S. Court of Appeals has ruled that the FCC does not have the authority to impose this methodology on the states,[9] which are ultimately charged with setting the prices for unbundled network elements, many states are setting prices that essentially follow the TELRIC methodology.

The result of U.S. policy, established in the Telecommunications Act of 1996 and the FCC's orders implementing the Act, is that entrants may rely artificially on unbundled elements provided by the incumbents rather than the construction of competing facilities. The mandated unbundling interferes with the establishment of a competitive market, and the pricing methodology may retard or foreclose the development of infrastructure competition. The U.S. policy may actually serve to entrench monopoly in the networks of the incumbents. As such, the U.S. unbundling policy is actually anticompetitive. Indeed, what is surprising about the policy is that it appears to have been adopted without any consideration of U.S. antitrust

9. Hansen, Circuit Judge, "Opinion and Order," Iowa Utilities Board v. Federal Communications Commission, *et al.*, On Petitions for Review of an Order of the Federal Communications Commission, Before Bowman, Wollman and Hansen, Circuit Judges, filed on July 18, 1997.

principles or of the numerous U.S. court decisions through which the development of the essential facilities doctrine can be traced.

Canada, in contrast, has chosen a significantly different regulatory policy regarding unbundling and pricing. Mandatory unbundling in Canada is, with some exceptions, limited to essential facilities. These essential facilities are telephone numbers (central office codes), directory listings that all carriers including new entrants control, and local loops in high-cost areas that the incumbent companies offered to treat as essential facilities until more evidence became available. Furthermore, the incumbent companies offered to make available on an unbundled basis a limited number of public service and safety services including 911 emergency service and relay service. Although the Commission was of the opinion that making available certain other unbundled services to competitors, primarily loops in urban areas and transiting between entrants, might facilitate the introduction of competition, the Commission recognized that these services are not essential facilities. As such, then, the Commission ordered that they be made available only for 5 years (CRTC Local Competition Decision 97-8, 1997, paragraph 86).

This approach provides an opportunity for new entrants to enter and establish themselves in the market knowing that the availability of these non-essential facilities is only guaranteed for 5 years. The incentives for them to construct their own facilities are clearly present, and the incumbents' incentives for research and development are only minimally affected. This approach, therefore, avoids nearly all of the problems and anticompetitive consequences evident in the U.S. approach.

In terms of pricing unbundled elements, the Commission in Canada has also taken a more procompetitive approach. The prices for unbundled elements are based on the incumbent's actual incremental cost,[10] and they also include a markup of 25% to recover a portion of shared and common costs. By basing prices on actual incremental costs, any inefficiencies that may be built into the incumbent's network will be reflected in the costs, and competitors will be given a realistic signal as to the costs and prices against which they will need to compete. This approach avoids many of the pitfalls of the U.S. approach and does not create artificial incentives for competitors to purchase unbundled facilities from the incumbent carriers, incentives and opportunities that are inconsistent with competitive markets. Furthermore, from a more practical perspective, Canadians resisted the temptation to embark on the development of a brand new, untested, and undeveloped incremental cost standard. The CRTC stayed with the incremental cost standard it has developed and employed for almost 20 years. The advantages of

---

10. For the pricing of unbundled elements, the Commission used its "Phase II" approach. This is an approximation of long-run incremental cost, with some not-great differences resulting from the time period used for the study that may be, in some cases, too short to measure true LRIC, and from a small allocation of shared costs.

this, contrasted with extensive litigation in the United States over a new (TELRIC) cost standard, are immense.

Resale

The difference between the Canadian and U.S. approaches is also evident in the case of resale. The Telecommunications Act of 1996 mandated the resale of local telephone service (Telecommunications Act of 1996, Section 251(c)(4)), as has the CRTC's Decision in Canada (CRTC Local Competition Decision 97-8, Section XII). The simple act of mandating resale of local service itself results in some anticompetitive problems, but, as discussed earlier, the major anticompetitive issue surrounding the resale of local telephone service concerns mandated wholesale discounts. Here there is a significant and important difference between the approaches taken in the United States and in Canada.

The Telecommunications Act of 1996 requires that local telephone service be provided on a wholesale basis at a price that reflects the retail price charged by the incumbent less any costs that are avoided by providing the service on a wholesale basis rather than on a retail basis (Telecommunications Act of 1996, Section 252(d)(3)). Under some interpretations of this standard, any anticompetitive consequences might have been minor, but the FCC's local competition order makes this situation worse. The FCC has taken the Act's "avoided cost" standard and converted it into an "avoidable cost" standard.[11] The FCC's approach is a significant departure from the plain words of the Act. The Act contemplates reducing the retail price by the actual costs that are saved in providing wholesale service. There is also some expectation in the Act that any additional costs incurred by the incumbent in providing a wholesale service could be recovered in the price charged for the wholesale service, but the FCC ignored this rather basic concept as well, and the FCC's "avoidable" costs are calculated entirely differently.

The FCC wants to measure all of the costs that can be potentially saved if the incumbent provided only a wholesale service and offered no retail service. For example, under the FCC's avoidable cost methodology, an incumbent might have no marketing expenses for wholesale service, and thus all of the incumbent's marketing costs should be subtracted from the retail tariff to get the wholesale rate. In the real world, however, the incumbent will be providing both retail and wholesale service; marketing costs may not be reduced at all if the incumbent provides some wholesale service, so the avoided, as opposed to avoidable, costs may be zero in this case. In any case, it is likely that such costs do not decrease proportionately with an increase in wholesale business.

The FCC has further suggested that wholesale discounts of 17% to 25% are appropriate in the absence of specific cost studies.[12] Although the U.S. Court of Ap-

---

11. FCC 96-325, par. 911.
12. FCC 96-325, par. 910.

peals has ruled that the FCC cannot impose such pricing rules for local service on the states,[13] many states have imposed such discounts on the ILECs. These discounts, which are larger than most of the resale agreements that had been negotiated prior to the FCC issuing its order, are not supported by any cost data. The consequence of discounts of this size is that entering the local market by reselling the incumbent's service is economically more attractive than entering by constructing facilities, since such a margin is guaranteed. This result is anticompetitive: it will hinder, if not foreclose, the development of a competitive local telecommunications market based on competing network facilities and, like the unbundling provisions of the Act, will entrench monopoly in the incumbent's local network.

Canada's resale policy is in stark contrast to the U.S. policy. The Commission recognized that facilities-based competition would be necessary for a truly competitive local telecommunications market to emerge in Canada (CRTC Local Competition Decision 97-8, paragraph 237). The Commission rejected mandated wholesale discounts and followed the approach it had established in the toll market, only mandating resale at the retail tariff (CRTC Local Competition Decision 97-8, paragraphs 250 and 252). Although market forces may eventually result in avoided-cost-based, market-responsive wholesale discounts based on volume and term commitments, such as are beginning to appear in the toll market, the Commission has not ordered artificial (or regulatorily imposed) discounts.

CONCLUSIONS

There are clear differences between the local competition policies selected by regulators in the United States and in Canada. U.S. policies make facilities-based competition more difficult, and, as such, may also slow down any moves toward bundling and convergence. In addition, to the extent that U.S. policies slow down the growth of facilities-based local competition, it will also make it more difficult for the Bell operating companies (BOCs) to enter the in-region interLATA (local access and transport area) toll business, thereby limiting the degree of competition in the toll market as well. Furthermore, local service competitors will always be able to count on the advantage of competing against a regulated competitor. Canada, of course, with a policy that does not discriminate against facilities-based competition, will see more competition, more bundling, and more convergence.

These are important differences, and the competitive situation may evolve in substantially different ways in the two countries, even though there is currently more facilities-based competition in the United States than in Canada as a result

---

13. Hansen, Circuit Judge. "Opinion and Order." Iowa Utilities Board v. Federal Communications Commission, *et al.* On Petitions for Review of an Order of the Federal Communications Commission. Before Bowman, Wollman and Hansen, Circuit Judges. Filed on July 18, 1997.

of an earlier lowering of barriers to competition. Now, however, the incentives in Canada favor facilities-based local competition more strongly than do the incentives in the United States. There is also reason to expect that the Canadian policy can work. New Zealand, for example, without an industry regulator and without mandated resale at any price and without mandated unbundling, is witnessing the development of facilities-based competition for both local and toll service.

As the European Union and other countries around the world also strive to develop policies so that they, too, can benefit from increased competition in telecommunications, it will be worthwhile to pay particular attention to the Canadian approach to local competition policy. The policies in the United States may limit the extent of true facilities-based competition, but Canadian local competition policy avoids the illusion of competition from the rapid entry of resellers of the incumbent's underlying service. Canadian policy holds out for the real benefits of competition that can only be realized from the presence of facilities-based competitors. It is in this way that consumers in Canada will enjoy all of the benefits of competitive telecommunications markets. Europeans, too, can benefit from real competition if governments and their regulators resist the temptation to follow the Americans into policies of competition based on regulation and, instead, follow the Canadian example of developing competition policy based on the principles of competition law and economics.

ACKNOWLEDGMENTS

We both participated, on behalf of TELUS Corporation (the Alberta, Canada, incumbent telephone company) and Stentor (the affiliation of Canadian incumbent telephone companies) in the local competition proceeding before the Canadian Radio-Television and Telecommunications Commission which led to Decision 97-8 issued on May 1, 1997. We have also served as consultants to TELUS on a variety of related issues, and we have benefited from numerous discussions with TELUS employees and consultants. We also wish to thank James Smallwood and Mark Murakami for assistance with the preparation of this chapter.

REFERENCES

Areeda, P. (1989). Essential facilities: An epithet in need of limiting principles. *Antitrust Law Journal, 58*, 838 888.

Canadian Radio-Television and Telecommunications Commission. Local Competition. Telecom Decision 97-8 (1997).

Federal Communications Commission. Implementation of the Local Competition Provisions of the Telecommunications Act of 1996 and Interconnection between Local Exchange Carriers and Commercial Mobile Radio Service Providers. CC Docket No. 96-98 and CC Docket No. 96-185, August 8, 1996, First Report and Order, FCC 96-325.

Federal Communications Commission. Federal State Joint Board on Universal Service. CC Docket No. 96-45, May 7, 1997, Report and Order, FCC 97-157.

Federal Communications Commission. Access Charge Reform, Price Cap Performance Review for Local Exchange Carriers, Transport Rate Structure and Pricing, and End User Common Line Charges. CC Docket No. 96-262, CC Docket No. 94-1, CC Docket No. 91-213, and CC Docket No. 95-72, May 7, 1997, First Report and Order, FCC 97-158.

Grieve, W. A., & Levin, S. L. (1995). *Local exchange interconnection and network component unbundling*. Paper presented at the International Telecommunications Society Workshop on Interconnection, Wellington, New Zealand.

Grieve, W. A., & Levin, S. L. (1996). Common carriers, public utilities and competition. *Industrial and Corporate Change, 5*, 993 1011.

Hansen, Circuit Judge. Order Granting Stay Pending Judicial Review. Iowa Utilities Board v. Federal Communications Commission, *et al.* U. S. Court Of Appeals, Eighth Circuit, Before Bowman, Wollman, and Hansen, Circuit Judges. Filed on October 15, 1996.

Hansen, Circuit Judge. Opinion and Order. Iowa Utilities Board v. Federal Communications Commission, *et al.* On Petitions for Review of an Order of the Federal Communications Commission. Before Bowman, Wollman and Hansen, Circuit Judges. Filed on July 18, 1997.

Hatfield Associates, Inc. (1994). *The cost of basic universal service*. Boulder, CO: Hatfield Associates, Inc.

Telecommunications Act of 1996, Pub L. No. 104-104, 110 Stat. 56, codified at 47 U.S.C., 151 et. seq.

Author Index

Subject Index

A

Action for Children's Television (ACT), 118
Alaska, 249–250
America's Carriers Telecommunications Association (ACTA), 195
AM stereo
  FCC policy on
    and competition, 130–132
    and consumer demand, 129, 136–137, 147
    and market fragmentation, 129, 135–136, 147
    and Motorola system, 134–136, 140–142, 147
    and public interest, 129–130
    and standards, 133–135
    and technology, 132, 147
  study data
    audience behavior, 139–140
    market activity, 140–142
    radio station characteristics, 140
  study estimation, 142–144, 150–151
  study methodology, 138–139
  study results, 147–148, 152–153
Apstar, 235–237
Ashbacker Radio v. FCC, 93
Asia-Pacific region, *see* Satellite policy, Asia-Pacific region
Asiasat, 235–237
Australia, children's educational television
  and advertisers, 124
  age specificity, 123–124

Children's Australian Drama (CAD), 116–117
Children's Television Standards (CTS), 115–117
  classification criteria, 125–127
  current regulation, 116–117
  historical regulation, 113–116
  program quality, 124–125
  U.S. comparison, 112–113, 125–128
Australian Broadcasting Authority (ABA), 112–113, 116, 123–125
Australian Broadcasting Corporation (ABC), 113

B

Benchmark Cost Model (BCM), 21
Broadcasting Services Act (1992) (Australia), 116
By-pass traffic, *see* International interconnection routing

C

Call-back services
  accounting rate by-pass, 71
  call-back, 71–72
  call stimulation, 73, 80–81
  half circuit, 70
  and international accounting rate system, 69–71
  International Direct Dial (IDD), 69, 71–72, 75–76
  international simple resale (ISR), 71